QA
157
.B63
2003

CLEARI

Chicago Public Library

Forgotten algebra : a se...    efr

MAY    2004

# Forgotten Algebra

## A Self-Teaching Refresher Course

### (with the optional use of the graphing calculator)

**Third Edition**

**Barbara Lee Bleau, Ph.D.**
Special Assistant to the Vice President
and
Professor of Mathematics
Florida Keys Community College
Key West, Florida

**BARRON'S**

To Bill
Before there was you . . . I forget.

© Copyright 2003, 1994, 1983 by Barron's Educational Series, Inc.

All rights reserved.
No part of this book may be reproduced
in any form, by photostat, microfilm, xerography,
or any other means, or incorporated into any
information retrieval system, electronic or
mechanical, without the written permission
of the copyright owner.

*All inquiries should be addressed to:*
Barron's Educational Series, Inc.
250 Wireless Boulevard
Hauppauge, New York 11788
http://www.barronseduc.com

*Library of Congress Catalog Card No.* 2003043546
International Standard Book No. 0-7641-2008-5

**Library of Congress Cataloging-in-Publication Data**
Bleau, Barbara Lee.
    Forgotten algebra : a self-teaching refresher course (and the optional use of the graphing
calculator) / Barbara Lee Bleau.—3rd ed.
        p.   cm.
    Includes index.
    ISBN 0-7641-2008-5
    1.  Algebra–Problems, exercises, etc.   I. Title.

QA157. B63 2003
512'.0076–dc21

                                                                                        2003043546

PRINTED IN THE UNITED STATES OF AMERICA
9 8 7 6 5 4 3 2

# Contents

CHICAGO PUBLIC LIBRARY
CLEARING BRANCH
6423 WEST 63RD PLACE
CHICAGO, ILLINOIS 60638-5005

# Preface

Algebra is not difficult! It is the primary tool used in a variety of quantitative courses in mathematics, economics, and engineering. This teach-yourself text-workbook was designed to provide students regardless of their previous background with the means to attain or improve their proficiency in this fundamental skill.

*Forgotten Algebra* can supplement various traditional mathematics courses, or it can be used to "brush up" before entering a course or preparing for a standardized entrance examination such as the SAT, GRE, or GMAT. It can also serve as an excellent introduction for the student who has never studied algebra. For example, each unit provides comprehensive explanations plus numerous examples, problems, and exercises, which include detailed solutions to facilitate self-study. In addition, this edition has incorporated an optional section at the end of many units entitled "Algebra and the Calculator." These sections illustrate the use of graphing calculators but are not designed to replace the manual exercises. Explanations are provided for using the Texas Instruments TI-83 model, which is widely available at a modest price. Be circumspect in its use, however, since your time and energy are best served learning algebra rather than learning how to use a graphing calculator.

Remember, *Forgotten Algebra* is a fundamental self-help approach to providing a workbook that is easy to read yet a comprehensive guide to learning algebra.

January 2003                                               Barbara Lee Bleau McKinley, Ph.D.

# UNIT 1

## Signed Numbers

In this first unit of your workbook you will learn the meanings of the terms *signed number* and *absolute value*, and the rules for performing the four basic operations of addition, subtraction, multiplication, and division using signed numbers, and how to evaluate numerical expressions using order of operations.

We will start with the concept of a signed number. A **signed number** is a number preceded by a plus or minus sign. A number preceded by a minus sign is called a **negative number**. For example, $-13$ and $-\frac{2}{5}$ are negative numbers. A plus sign is used to denote a **positive number**; $+5$ is a positive number. If no sign is written, the number is understood to be positive; 20 is a positive number. Zero is the only exception. The number 0 is neither positive nor negative.

On a number line (think of a thermometer on its side), all the numbers to the right of 0 are positive numbers and all the numbers to the left of 0 are negative numbers. The number 0, often referred to as the origin, is neither positive nor negative.

Next is the concept of absolute value. The definition that I especially like is that the **absolute value** of a signed number is the number that remains when the sign is removed. For example, the absolute value of $-14$ is 14; when the negative sign is removed, the number that remains is 14. Likewise, the absolute value of $+2$ is 2.

Instead of writing the words *absolute value*, we can use the symbol $||$. Thus, $|-14|$ is read as "the absolute value of negative 14." And $|-14| = 14$.

Geometrically absolute value can be defined as the distance on the number line from a number to 0 without regard to direction. For example, the distance on the number line from 4 to 0 is 4 units, and the distance from $-4$ to 0 is 4 units. Thus, $|4| = 4$ and $|-4| = 4$.

Before continuing, here are a few problems for you to try.

---

Problem 1     Find the absolute value of –37.

Problem 2     Find |+5|.

Problem 3     Find |–11|.

Problem 4     Find |0|.

Answers:    1. 37     2. 5     3. 11     4. 0

---

# ADDING SIGNED NUMBERS

Rule 1:     **To add numbers with like signs, add their absolute values. To this result prefix the common sign.**

---

## EXAMPLE 1    (–5) + (–3)

---

Solution:    The absolute value of (–5) is 5, or |–5| = 5, and the absolute value of (–3) is 3, or |–3| = 3. Add the absolute values 5 and 3. To the result, 8, prefix the common sign. Thus (–5) + (–3) = –8.

---

## EXAMPLE 2    (+5) + (+3)

---

Solution:    Use the same reasoning as for Example 1, but the common sign now is plus. Thus, (+5) + (+3) = +8.

---

Rule 2:     **To add two numbers with unlike signs, subtract the smaller absolute value from the other. To this result, prefix the sign of the number having the larger absolute value.**

## EXAMPLE 3   $(-5) + (+3)$

Solution:   $|-5| = 5$, and $|+3| = 3$. Subtract the smaller absolute value, 3, from the larger absolute value, 5. To the result, 2, prefix the sign of the number having the larger absolute value, in this example, 5. Thus $(-5) + (+3) = -2$.

Now you try a few.

Problem  5     $(+5) + (+7)$

Problem  6     $(-9) + (-2)$

Problem  7     $(-1) + (-2) + (-3)$

Problem  8     $(-10) + (+2)$

Problem  9     $(7) + (-3)$

Problem 10     $(-2) + (+8)$

Problem 11     $-1 + 7$

Problem 12     $12 + (-1)$

Answers:   5. 12     6. −11     7. −6     8. −8
9. 4     10. 6     11. 6     12. 11

I hope you noticed that, when adding two signed numbers, you add absolute values for like signs and subtract absolute values for unlike signs.

I fully realize you could have gotten all of Problems 5–12 correct, without ever knowing the two rules, by simply using a calculator. And if the numbers had been more complicated, say $(-1237.15) + (352.79)$, I expect that's what I would have done—used my calculator. But for easy numbers, like those above, it's faster to do them in your head than to rely on a calculator.

## SUBTRACTING SIGNED NUMBERS

Rule 3:   **To subtract a signed number, add its opposite.**

## EXAMPLE 4   $(5) - (-2)$

Solution:   To subtract −2 from 5, the rule says to add the opposite of −2. The opposite of −2 is 2.
Thus, $(5) - (-2) = 5 + 2 = 7$.

## EXAMPLE 5 $(-6) - (-2)$

Solution: The opposite of $-2$ is $2$.
Thus, $(-6) - (-2) = -6 + 2 = -4$.

## EXAMPLE 6 $(-3) - 6$

Solution: The opposite of $6$ is $-6$.
Thus, $(-3) - 6 = (-3) + (-6) = -9$.

In each of the above examples, the subtraction was changed to addition by rewriting the problem with the signed number's opposite. Then the final answer was found by using the rules for the addition of signed numbers.

Here are some problems for you. Try them without using a calculator.

| Problem 13 | $23 - 18$ |
| Problem 14 | $5 - (17)$ |
| Problem 15 | $(-3) - (2)$ |
| Problem 16 | $2 - (-7)$ |
| Problem 17 | $-10 - (-5)$ |
| Problem 18 | $-11 - 2$ |

Answers: 13. 5    14. −12    15. −5    16. 9    17. −5    18. −13

# MULTIPLYING SIGNED NUMBERS

Rule 4:    **To multiply two signed numbers with like signs, multiply their absolute values and make the product positive.**

Rule 5:    **To multiply two signed numbers with unlike signs, multiply their absolute values and make the product negative.**

## EXAMPLE 7 $(5)(2) = 10$

## EXAMPLE 8 $(-3)(-5) = +15$

## EXAMPLE 9 $(-1)(+7) = -7$

## EXAMPLE 10 $(+3)(-2) = -6$

Think of it like this: Two like signs yield plus, and two unlike signs yield minus. Now try the following problems.

---

Problem 19     (–3)(5)

Problem 20     (–2)(–2)

Problem 21     (–4)(–11)

Problem 22     6(–8)

Problem 23     (–1)(7)

Problem 24     (4)(–3)

Answers:  19. –15     20. 4     21. 44
22. –48     23. –7     24. –12

---

## DIVIDING SIGNED NUMBERS

Rule 6:   **To divide two signed numbers with like signs, divide their absolute values and make the quotient positive.**

Rule 7:   **To divide two signed numbers with unlike signs, divide their absolute values and make the quotient negative.**

---

EXAMPLE 11     $-35/5 = -7$

EXAMPLE 12     $-63/-7 = 9$

EXAMPLE 13     $10/-5 = -2$

EXAMPLE 14     $-12/-2 = 6$

---

## ORDER OF OPERATIONS

By now you should be asking yourself, What happens when the problem involves more than one operation? In such cases we use the following order of operations.

Step 1:   Working from left to right, do any multiplications or divisions in the order in which they occur.

Step 2:   Again working from left to right, do any additions or subtractions in the order in which they occur.

## EXAMPLE 15

Calculate:   $3 + 8 \div 4$.

---

Solution:   $3 + 8 \div 4 = 3 + \mathbf{2}$     Divide

$= 5$          Add

---

## EXAMPLE 16

Calculate:   $16 \div 2 \cdot 4 - 6 \cdot 2$.

---

Solution:   $16 \div 2 \cdot 4 - 6 \cdot 2 = \mathbf{8} \cdot 4 - 6 \cdot 2$     Divide

$= \mathbf{32} - \mathbf{12}$     Multiply

$= 20$          Subtract

---

In each of the above examples, I started on the left and did the multiplications and divisions in the order they occurred. Then, returning to the left, all the additions and subtractions were calculated in the order they occurred.

As problems increase in complexity, we will add additional steps; but for now practice using these two.

---

**Problem 25**   Calculate:   $5 + 2 \cdot 4 - 3$.

**Problem 26**   Calculate:   $27 \div 9 - 2 \cdot 7$.

Answers:   25. 10   26. −11

---

Algebra is an essential branch of mathematics. As such, it is important that you learn to do the math manually, that is, with paper and pencil. Notwithstanding, graphing calculators have become a way of life for some people, and I want to suggest that we take advantage of the technology available to us. All the material in this text is designed to be done manually. But at the end of some units there is an optional section entitled "Algebra and the Calculator." Explanations are provided for using a TI-83 graphing calculator. I use mine mostly when the numerical calculations are too "messy" to be done by hand and for checking my work. Optional exercises for practice and solutions are included as well. Remember, though, I want you to spend your time and energy learning algebra, which is the focus of this book, rather than learning how to use a graphing calculator.

## ALGEBRA AND THE CALCULATOR (Optional)

By now maybe you've decided that a calculator is the way to go rather than learning all these rules. In each of the following examples, I have shown the keystrokes necessary to solve the problem with a calculator. Until you are more familiar with the rules of algebra it is important to enter the problem exactly as shown, especially the parentheses and signs.

---

**EXAMPLE 17**    Using a calculator, find:    −2 + (−9).

Solution:    To find the answer, press ⌊(−)⌋ ⌊2⌋ ⌊+⌋ ⌊(⌋ ⌊(−)⌋ ⌊9⌋ ⌊)⌋ ⌊ENTER⌋. Note: The ⌊(−)⌋ key is used to denote a negative number. An error message will appear if the subtraction key ⌊−⌋ is used by mistake.

Answer:    −11

---

**EXAMPLE 18**    Using a calculator, find:    |−6| + 3.

Solution:    Press ⌊MATH⌋, select NUM, select 1:abs, and press ⌊ENTER⌋. Press ⌊(−)⌋ ⌊6⌋ ⌊)⌋ ⌊+⌋ ⌊3⌋ ⌊ENTER⌋.

Notice the number of keystrokes needed to do this problem. You will save yourself a great deal of time if you learn to recognize problems like this as nothing more than asking for the sum: 6 + 3 = 9.

Answer:    9

---

You should now have an understanding of signed numbers, absolute value, and the order of operations, and be able to add, subtract, multiply, and divide signed numbers. Briefly stated, the rules are as follows:

**Addition:**    When the signs are the same, add the absolute values and keep the common sign. With unlike signs, subtract the absolute values and keep the sign of the number with the larger absolute value. (Rules 1 and 2)

**Subtraction:** Change subtraction to addition of the opposite, and follow the rules for addition. (Rule 3)

**Multiplication and division:** When the signs are the same, multiply or divide the absolute values and make the answer positive. With unlike signs, multiply or divide the absolute values and make the answer negative. (Rules 4–7)

Before beginning the next unit, try the following exercises without using a calculator. When you have finished, check your answers against those at the back of the book.

# EXERCISES

Perform each of the indicated operations.

1. $-2 + (-5)$
2. $1 - (-3)$
3. $8 - 17$
4. $-4 - (-3)$
5. $6(-8)$
6. $(-3)(6)$
7. $-4(-11)$
8. $32/-8$
9. $-25/-5$
10. $-3 + |-4|$
11. $|-6| - 7$
12. $90/-30$
13. $-11 + 20$
14. $-6 - (-3)$
15. $-3 + 2 + (-5)$
16. $-5 - 5$
17. $(-2)(-3)(-4)$
18. $(-4)(-8)$
19. $-3 + 1 - (-2)$
20. $(-1)(-1)(-1)(-1)$
21. $1 + 6 \cdot 3 \div 2 - 2$
22. $12 \div 3 \cdot 2 - 1$
23. $0 \cdot 7 + 15 \div 5 - 2 \cdot 1$
24. $(-10) \cdot 12 - 6 \cdot 3$
25. $|-5| + |3| - |-2|$
26. $-|-3| + |7|$

# UNIT 2

## Grouping Symbols and Their Removal

In this unit you will learn the meanings of some of the words we will be using throughout the course. Then you will learn about various grouping symbols and the way they are removed to simplify expressions.

## SOME IMPORTANT DEFINITIONS

Five of the most important concepts you will need in the course are: variable, term, coefficient, expression, and like terms.

Here are some "commonsense" or "working" definitions of these concepts. It is *not* important that you memorize these definitions, but you should understand each one and be able to define the concept in your own words.

A **variable** is a letter or symbol used to represent some unknown quantity. For example, $x$ is often used as a variable.

A **term** is a symbol or group of symbols separated from other symbols by a plus or minus sign. For example, in $3x - 5y + 7xyz - 2$, there are four terms. Note that terms can contain a number and/or several letters. Therefore $7xyz$ is a single term; the number is written first, and the letters are usually put in alphabetical order.

The numerical **coefficient** of a variable is the number that multiplies the variable. In the term $2y$ the coefficient of $y$ is 2. Here are some examples:

a. $8w$    The numerical coefficient of $w$ is 8.

b. $x$    When no number is written, the numerical coefficient is understood to be 1.

c. $-5z$    The numerical coefficient is $-5$.

The sum or difference of one or more terms can be referred to as an **expression**. Expressions are given different names depending on the number of terms involved. Expressions with one term are called monomials, whereas expressions with two terms are called binomials. The expression $3x - 5y + 4w$ has three terms and is called a trinomial.

**Like terms** contain the same variable or variables and differ only in their numerical coefficients. In the expression $3x + 4x$ the terms are like terms that differ only with respect to their coefficients, 3 and 4. Like terms can be combined. The terms of the expression $3x + 4x$ can be combined to form a single term, $7x$. Similarly, like terms in the trinomial expression $3x - 5y + 4x$ can be combined, forming a binomial expression, $7x - 5y$.

Try a few problems to test your understanding of these important concepts.

---

| | | |
|---|---|---|
| Problem 1 | How many terms are there? | $10zy + 3x - 5an + 12$ |
| Problem 2 | Which is the *variable*? | $7a$ |
| Problem 3 | What is the *coefficient* of y? | $5y$ |
| Problem 4 | A binomial expression has _____ terms. | |
| Problem 5 | When the like terms in the expression $10y + 8z + 3y$ are combined, the result will be _____. | |

Answers:  1. four  2. *a*  3. 5  4. two  5. $13y + 8z$

---

# GROUPING SYMBOLS

Now let's turn our attention to grouping symbols. There are three common types of grouping symbols:

$$\begin{array}{ll} \text{parentheses} & (\ ) \\ \text{brackets} & [\ ] \\ \text{braces} & \{\ \} \end{array}$$

Whether you use parentheses, brackets, or braces, these symbols indicate that the quantities enclosed within them are considered to be a single unit with respect to anything outside the grouping symbol. In the expression $8y + (3a - b) + 2x$, the terms $3a - b$ are considered as a single unit, separate from the terms outside the parentheses.

## REMOVING GROUPING SYMBOLS AND SIMPLIFYING

To remove grouping symbols we use the **distributive property:** $a(b + c) = ab + ac$. The distributive property tells us to remove parentheses by multiplying each term inside the parentheses by the term in front of the parentheses.

**EXAMPLE 1**   $2(x + y) = 2x + 2y$

In an expression like $2(x + y)$, we multiply each term inside the parentheses by 2.

**EXAMPLE 2**   $-3(x - 5) = -3x + 15$

Note the plus sign between the terms since $(-3)(-5) = +15$.

_____

Next we consider some more complicated expressions involving like terms. To **simplify** these expressions, we follow the rules of order of operations from Unit 1. Working from left to right, first remove the grouping symbols (step 1, do any multiplications or divisions) followed by combining any like terms (step 2, do any additions and subtractions).

## EXAMPLE 3

Simplify:   $2(4xy + 3z) + 3[x - 2xy] - 4\{z - 2xy\}$.

_____

Solution:   $2(4xy + 3z) + 3[x - 2xy] - 4\{z - 2xy\}$

$= 8xy + 6z + 3x - 6xy - 4z + 8xy$

Note plus sign.

$= 3x + 2z + 10xy$

_____

## EXAMPLE 4

Simplify:   $3aw + (aw + 4z) - (2x - 3y + 4z)$.

_____

Solution:   $3aw + (aw + 4z) - (2x - 3y + 4z)$

$= 3aw + aw + 4z - 2x + 3y - 4z$

Note: To remove the first set of parentheses we are actually multiplying by 1, which is understood to be there but is not written. In an expression like $-(2x - 3y + 4z)$, remove the parentheses by multiplying each term inside the parentheses by $-1$. Again the 1 is not written, but it is understood to be there.

$= 4aw - 2x + 3y$

_____

Try the next two on your own.

## Problem 6

Remove the grouping symbols and simplify: $3\{x - 10 + 2y\} - 2$.

Solution:

Answer:   $3x + 6y - 32$

## Problem 7

Simplify:   $2(x - 1) - 3(2 - 3x) - (x + 1)$.

Solution:

Answer:   $10x - 9$

Occasionally more complicated expressions occur with parentheses within parentheses. When one set of symbols is within another, the grouping symbols must be removed from the inside out. One example follows.

## EXAMPLE 5

Simplify:   $x - [5 - 2(x - 1)]$.

Solution:   $x - [5 - 2(x - 1)]$

$= x - [5 - 2x + 2]$
$= x - [5 - 2x + 2]$

$= x - 5 + 2x - 2$
$= 3x - 7$

You should now be able to define, in your own words, the basic concepts *term*, *variable*, *coefficient*, *expression*, and *like terms*. These basic concepts will be used continually throughout the following units.

You should also be familiar with three common types of grouping symbols. Remember that the quantities enclosed within these symbols are considered to be a single unit, separate from anything outside the parentheses, brackets, or braces.

Remember also that, when one set of symbols is within the other, grouping symbols are removed by working from the inside out. Simplifying then involves combining the like terms.

Before beginning the next unit you should simplify the following expressions by removing the grouping symbols and collecting like terms. When you have completed them, check your answers against those at the back of the book.

## EXERCISES

Simplify:

1. $-7(x + 2y - 3)$
2. $3x + 4x + (x + 2)$
3. $2(x + y) + 7x + 3$
4. $-(-2x + 1) + 1$
5. $7x - 2y + 5 + (2x + 5y - 4)$
6. $(3x + 5xy + 2y) + (4 - 3xy - 2x)$
7. $(5y - 2a + 1) + 2(y - 3a - 7)$
8. $3(2a - b) - 4(b - 2a)$
9. $2(7x - 5 + y) - (y + 7)$
10. $5 - 2(x + 2[3 + x])$
11. $x - \{2x - 2(1 - x)\}$
12. $4 - 9(2x - 3) + 7(x - 1)$
13. $3(x + 4) + 5x - 8$
14. $4x - [5 - 3(2x - 6)]$

# UNIT 3

## Solving First-Degree Equations

The purpose of this unit is to provide you with an understanding of first-degree equations. When you have finished the unit, you will be able to identify first-degree equations, distinguish them from other types of equations, and solve them.

What is a first-degree equation? A **first-degree equation** has these characteristics:

1. There is **only one variable**.

2. The variable is involved in **one or more of only the four fundamental operations** of addition, subtraction, multiplication, and division.

3. The variable is **never multiplied by itself**.

4. The variable does **not** appear **in any denominator**.

Here are some examples of first-degree equations:

$$2x + 3 = 15$$
$$\sqrt{5}x - \frac{x+2}{2} = \pi$$
$$3(x - 1) + 2 = 0$$

Here are some examples that are *not* first-degree equations:

$$x + 3y = 5$$
$$x^2 = 9$$
$$3 + \sqrt{x} = 2(x - 7)$$
$$\frac{2}{x} = 7(x - 1)$$
$$(x - 3)(x + 3) = 12$$

Before proceeding, determine why each of the above is *not* a first-degree equation.

Now we are ready to begin solving first-degree equations. You probably remember the basic strategy—isolate the variable on one side of the equal sign and get all other terms on the other side. To accomplish this, we use two rules. The first is:

Rule 1:   **A term can be transposed or moved from one side of an equation to the other if and only if its sign is changed to its opposite as it crosses the equal sign.**

When we transpose a negative term, we sometimes say that we "add the same number on both sides of the equation." Likewise, when we transpose a positive term, we can say that we "subtract the same number on both sides of the equation."

EXAMPLE 1     If     $x - 3 = 5,$

then          $x = 5 + 3,$

and          $x = 8.$

Note that the sign is changed.

EXAMPLE 2     If   $2 - x = 7,$

then   $2 - x = 7,$

$2 - 7 = x,$

and          $-5 = x.$

The second rule is:

Rule 2:   **Both sides of an equation can be multiplied or divided by the same nonzero number.**

EXAMPLE 3     If     $3x = 2,$

then          $\dfrac{3x}{3} = \dfrac{2}{3},$

$\dfrac{3x}{3} = \dfrac{2}{3},$

and          $x = \dfrac{2}{3}.$

EXAMPLE 4     If     $\dfrac{x}{4} = 12,$

then     $4\left(\dfrac{x}{4}\right) = 4(12),$

$\dfrac{4x}{4} = 48,$

and          $x = 48.$

Now let's get on with the business of solving equations.

> Definition: A **solution** of an equation is any number that makes the equation true when that number is substituted for the variable. Sometimes it is called the **root** of the equation.

To solve first-degree equations, I use a four-step strategy that can be used to solve *all* first-degree equations. The four steps are as follows:

> **Strategy for Solving First-Degree Equations**
>
> 1. **Simplify:**
>    a. Remove parentheses.
>    b. Clear any fractions.
>    c. Collect like terms.
>
> 2. **Transpose:**
>    Isolate all terms with the variable to one side and transpose everything else to the other side. Remember to change the term's sign when crossing the equal sign.
>
> 3. **Simplify.**
>
> 4. **Divide** each term of the entire equation **by the coefficient** of the variable.

Here are some examples that illustrate the use of these steps in solving first-degree equations. Read through each step, and be sure you understand what has happened to the terms.

# EXAMPLE 5

Solve the equation for $x$:  $5x + 10 - 3x = 6 - 4x + 16$.

Solution:

$$5x + 10 - 3x = 6 - 4x + 16$$

1. **Simplify** by collecting like terms.

$$2x + 10 = 22 - 4x$$

2. **Transpose.**

$$2x + 10 = 22 - 4x$$

$$2x + 4x = 22 - 10$$

3. **Simplify** by collecting like terms.

$$6x = 12$$

4. **Divide** by the coefficient of $x$.

$$\frac{6x}{6} = \frac{12}{6}$$

$$\frac{6x}{6} = 2$$

Answer:

$$x = 2$$

# EXAMPLE 6

Solve the equation for $x$:   $3(2x + 5) = 4x + 23$.

Solution: $$3(2x + 5) = 4x + 23$$

1. **Simplify** by removing parentheses. $$6x + 15 = 4x + 23$$

2. **Transpose.** $$6x + 15 = 4x + 23$$

$$6x - 4x = 23 - 15$$

3. **Simplify** by collecting like terms. $$2x = 8$$

4. **Divide** by the coefficient of $x$. $$\frac{2x}{2} = \frac{8}{2}$$

Answer: $$x = 4$$

# EXAMPLE 7

Solve the equation for $x$:   $4 - (2x - 8) = x$.

Solution: $$4 - (2x - 8) = x$$

1. **Simplify** by removing parentheses, and collecting like terms. $$4 - 2x + 8 = x$$
$$12 - 2x = x$$

2. **Transpose.** $$12 - 2x = x$$
$$12 = x + 2x$$

3. **Simplify** by collecting like terms. $$12 = 3x$$

4. **Divide** by the coefficient. $$\frac{12}{3} = \frac{3x}{3}$$

Answer: $$4 = x$$

Try to solve this first-degree equation yourself. Cover the solution below, and refer to it only after you have arrived at *your* solution.

## Problem 1

Solve the equation for $x$:           $x - 3 - 2(6 - 2x) = 2(2x - 5)$.

   1.  **Simplify:**   remove parentheses,     _____ = _____

                   collect like terms.     _____ = _____

   2.  **Transpose.**                                  _____ = _____

   3.  **Simplify:**   collect like terms.     _____ = _____

---

Solution:   $x - 3 - 2(6 - 2x) = 2(2x - 5)$

$$x - 3 - 12 + 4x = 4x - 10$$

$$5x - 15 = 4x - 10$$

$$5x - 4x = -10 + 15$$

$$x = 5$$

---

Now try solving this equation without any clues.

---

## Problem 2

Solve the equation for $x$:   $3x - 2(x + 1) = 5x - 6$.

---

Solution:

Answer: $x = 1$

---

    Many people seem to panic when fractions occur in a problem. For that reason, when attempting to solve an equation containing fractions, I suggest immediately multiplying the entire equation (Rule 2) by the common denominator to clear it of fractions. The next two examples will illustrate this concept.

## EXAMPLE 8

Solve the equation for $x$:   $\dfrac{3x}{2} - 5x = 6$.

Solution:

$$\dfrac{3x}{2} - 5x = 6$$

1.  **Simplify** by clearing fractions.

    (multiply equation by the denominator)

    $$2\left(\dfrac{3x}{2} - 5x = 6\right)$$

    (this cancels the denominator)

    $$\dfrac{2 \cdot 3x}{2} - 2 \cdot 5x = 2 \cdot 6$$

    (equation cleared of fraction)

    $$3x - 10x = 12$$

3.  **Simplify** by collecting like terms.

    $$-7x = 12$$

4.  **Divide by the coefficient of $x$.**

    $$\dfrac{-7x}{-7} = \dfrac{12}{-7}$$

Answer:

$$x = \dfrac{12}{-7} = \dfrac{-12}{7} = -\dfrac{12}{7}$$

Note:   The answer can be written in three different ways. I happen to prefer the second fraction with the negative in the numerator, but the final choice is yours.

## EXAMPLE 9

Solve:   $\dfrac{2}{5}x - 4 = \dfrac{1}{3}x + 1$.

Solution:

$$\dfrac{2}{5}x - 4 = \dfrac{1}{3}x + 1$$

1.  **Simplify** by clearing fractions.

    (multiply by the common denominator)

    $$15\left(\dfrac{2}{5}x - 4 = \dfrac{1}{3}x + 1\right)$$

    (this cancels the denominators)

    $$15 \cdot \dfrac{2}{5}x - 15 \cdot 4 = 15 \cdot \dfrac{1}{3}x + 15 \cdot 1$$

    (equation cleared of fractions)

    $$3 \cdot 2x - 60 = 5 \cdot 1x + 15$$
    $$6x - 60 = 5x + 15$$

2.  **Transpose.**

    $$6x - 60 = 5x + 15$$
    $$6x - 5x = 15 + 60$$

3.  **Simplify** by collecting like terms.

    $$x = 75$$

Answer:

$$x = 75$$

# TYPES OF EQUATIONS

Thus far, all the equations we have considered have been what are called conditional equations. A **conditional equation** is one that is true for only certain values of a variable. For example, $x + 5 = 8$ is true only for $x = 3$. Therefore $x + 5 = 8$ is a conditional equation.

Now in contrast, consider the following example.

## EXAMPLE 10

Solve:   $3x + 1 + x = 2(x + 1) + 2x - 1$.

Solution:                                                        $3x + 1 + x = 2(x + 1) + 2x - 1$

1. **Simplify** by removing parentheses,               $3x + 1 + x = 2x + 2 + 2x - 1$
   and collecting like terms.                                    $4x + 1 = 4x + 1$

2. **Transpose.**                                                    $4x + 1 = 4x + 1$

                                                                            $4x - 4x = 1 - 1$

3. **Simplify** by collecting like terms.                          $0 = 0$

Answer:   The solution is the entire set of real numbers.

Before continuing, let me make sure you understand why the solution to Example 10 is the set of real numbers. The final equation, $0 = 0$, is *always* true, regardless of the value of $x$, because 0 always equals 0. This type of equation is called an **identity**. Its solution is the entire set of real numbers. In other words, $x$ can equal any number.

For example, if $x = 5$, then $3(5) + 1 + 5 = 2(5 + 1) + 2(5) - 1$,

$$15 + 1 + 5 = 12 + 10 - 1,$$

$$\text{and } 21 = 21.$$

Or,              if $x = -3$, then $3(-3) + 1 + (-3) = 2(-3 + 1) + 2(-3) - 1$,

$$-9 + 1 + (-3) = -6 + 2 - 6 - 1,$$

$$\text{and } -11 = -11.$$

Try *any* value you like for $x$. Prove to yourself that it does indeed satisfy the equation.

Before leaving this example, notice that midway through the solution the left side of the equation was simplified to $4x + 1$, as was the right side of the equation. We could have stopped there because $4x + 1 = 4x + 1$ is *always* true, regardless of the value of $x$.

Solve the next problem on your own.

## Problem 3

Solve:    $\dfrac{3x}{2} - 10 = 6x.$

Solution:

Answer: $x = \dfrac{-20}{9}$

Will there always be a solution to an equation? What do you think? The answer is no, although the majority of the problems in this book have solutions because you need practice in solving equations. The last example illustrates how you can recognize when there is no solution to an equation.

## EXAMPLE 11

Solve:   $3(x + 1) + x = 4(x + 1).$

Solution:

$$3(x + 1) + x = 4(x + 1)$$

1. **Simplify** by removing parentheses,    $3x + 3 + x = 4x + 4$

   then collecting like terms.    $4x + 3 = 4x + 4$

2. **Transpose.**    $4x (+ 3) = (4x) + 4$

   $$4x - 4x = 4 - 3$$

3. **Simplify.**    $0 = 1$

Answer:   No solution

The final equation, $0 = 1$, is *always* false, regardless of the value of $x$, because 0 never equals 1. No matter what value is substituted in for the variable, the final statement will be false. Thus, whenever the concluding equation is false, there is no solution to the problem.

These same four steps apply when we solve first-degree equations of greater difficulty. For example, consider the equation: $3\{1 + 4x - (x + 1)\} = 0.$

Note that one pair of symbols is "nested" within the outer pair. Recall from the previous unit that we start by removing the innermost pair and then work outward. We will conclude this unit by your solving this last problem.

## Problem 4

Solve: $3\{1 + 4x - (x + 1)\} = 0$.

---

Solution:

Answer: $x = 0$

---

You should now be able to solve any first-degree equation. Remember that the basic strategy for solving a first-degree equation involves four steps:

1. **Simplify** by removing parentheses, clearing fractions, and then collecting like terms.

2. **Transpose.**

3. **Simplify** by collecting like terms.

4. **Divide by the coefficient of the variable.**

Also remember that the sign of a term changes when the term is moved across the equal sign, and that both sides of an equation can be multiplied or divided by the same nonzero number.

I believe this is one of the most important units of the book because first-degree equations reoccur throughout algebra. Therefore I want to encourage you to make sure you are able to successfully solve the following equations before beginning the next unit.

## EXERCISES

Solve:

1. $2x - 7 = 9 - 6x$

2. $2(x + 1) - 3(4x - 2) = 6x$

3. $\dfrac{x - 3}{4} = 5$

4. $20 - \dfrac{3x}{5} = x - 12$

5. $2x - 9 = 5x - 15$

6. $2(x + 2) = 5 + \dfrac{x + 1}{3}$

7. $15 - 3(9 - x) = x$

8.  $3 - \dfrac{5(x-1)}{2} = x$

9.  $x - \dfrac{x-1}{4} = 0$

10.  $1 - \dfrac{x}{2} = 5$

11.  $-5w - 1 = -9w - 1$

12.  $3(-2x + 1) = -6x - 7$

13.  $2(4a + 1) = 4(2a - 1) + 6$

14.  $3(z + 5) - 2z = \dfrac{z-1}{2} + 17$

15.  $13 - (2y + 2) = 2(y + 2) + 3y$

16.  $3x + 4(x - 2) = x - 5 + 3(2x - 1)$

17.  $2c + 3(c + 2) = 5c + 11$

18.  $2\{2 - x - (2x - 5)\} = 11 - x$

—————————————————————————————————————

# UNIT 4

## A Special Type of Equation: The Fractional Equation

In this unit you will learn about a special type of equation, the fractional equation, and the strategy used to solve it. When you have completed the unit, you will be able to identify and solve fractional equations.

> Definition:   A **fractional equation** is an equation in which the variable appears in a denominator.

For example, $\dfrac{2+x}{x} = 3$ is a fractional equation.

The same four steps we have been using to solve first-degree equations will continue to be used. However, a fifth step must be added when solving fractional equations. We must check to see whether our solution satisfies the **original** equation.

Recall that a solution is any number that makes the equation true when that number is substituted for the variable. Unfortunately there are several operations that may produce an equation not equivalent to the original equation. One of these operations is multiplying or dividing both sides of an equation by an expression containing the variable, which often occurs when solving a fractional equation. Thus, the final test of whether a number is part of the solution for a fractional equation is to insert it into the original equation and see whether it yields a true statement.

---

**Strategy for Solving Fractional Equations**

1. **Simplify:** remove parentheses,
   clear any fractions,
   collect like terms.

2. **Transpose.**

3. **Simplify.**

4. **Divide by the coefficient of the variable.**

5. **Check** by substituting the tentative answer into the original equation.

---

Keep in mind that as the equations become more complex, you might want to interchange the order in which you simplify. For example, it might be more logical in a given problem to clear the fractions before removing the parentheses. As long as you have carried out the algebraic procedures correctly, the resulting equations will be equivalent.

At this time it is necessary to remind you that **division by zero is undefined**. In other words, zero can never be the denominator of a fraction; $\frac{a}{0}$, for example, is meaningless.

We will now use the four basic steps and the check by substitution to solve our example of a fractional equation.

---

## EXAMPLE 1

Solve for $x$: $\quad \dfrac{2+x}{x} = 3.$

- - - - - - - - - - - - - - - - - - - - - - - - - - - - - - - - - - - - - - - -

Solution: 

1. **Simplify** by clearing fractions.

$$\left( \frac{2+x}{x} = 3 \right) x$$

$$\frac{(2+x)}{x} \cdot x = 3 \cdot x$$

$$\frac{(2+x)}{\cancel{x}} \cdot \cancel{x} = 3 \cdot x$$

$$2 + x = 3x$$

2. **Transpose.**

$$2 = 3x - x$$

3. **Simplify.**

$$2 = 2x$$

4. **Divide** by the coefficient of $x$.

$$\frac{\cancel{2}}{\cancel{2}} = \frac{\cancel{2}x}{\cancel{2}}$$

$$1 = x$$

5. **Check:** substitute the tentative answer in the **original** equation.

$$\frac{2+x}{x} = 3$$

$$\frac{2+1}{1} \overset{?}{=} 3$$

$$3 = 3$$

Answer: This is true. So $x = 1$ is the solution.

---

## EXAMPLE 2

Solve for $x$: $\dfrac{7}{x} + 3 = \dfrac{-1}{2}$.

---

Solution: What shall we use as the common denominator to clear the equation of fractions: $x$, $2x$, $3x$, $2$, etc.?

Using our method, we can make a wrong choice and still solve the equation simply by adding an extra step or two.

Suppose we use $x$:

$$\frac{7}{x} + 3 = \frac{-1}{2}$$

$$\left(\frac{7}{x} + 3 = \frac{-1}{2}\right)x$$

$$\frac{7}{x} \cdot x + 3 \cdot x = \frac{-1}{2} \cdot x$$

$$7 + 3x = \frac{-x}{2}$$

Obviously $x$ was not a wise choice as it did not clear the equation completely of fractions. However, it is easy to continue the solution.

Because there is still a denominator of 2, we can simply repeat the process using 2:

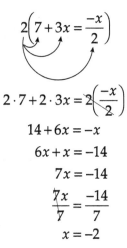

$$2 \cdot 7 + 2 \cdot 3x = 2\left(\frac{-x}{2}\right)$$

$$14 + 6x = -x$$

$$6x + x = -14$$

$$7x = -14$$

$$\frac{7x}{7} = \frac{-14}{7}$$

$$x = -2$$

Had we multiplied by $2x$ in the first step, the problem would have been much shorter.

Now we must check the tentative solution, using the original equation.

$$\frac{7}{x} + 3 = \frac{-1}{2}$$

$$\frac{7}{-2} + 3 \stackrel{?}{=} \frac{-1}{2}$$

$$-3\frac{1}{2} + 3 \stackrel{?}{=} -\frac{1}{2}$$

$$-\frac{1}{2} = -\frac{1}{2}$$

Answer:  This is true. So $x = -2$ is the solution.

- - - - - - - - - - - - - - - - - - - - - - - - - - - - - - - - - - - - - - - - - - - - -

Now you try one.

---

## Problem 1

Solve for $x$:  $\dfrac{2}{x+1} + 3 = 1$.

Hint:  $(x + 1)$ is the **entire** denominator.

- - - - - - - - - - - - - - - - - - - - - - - - - - - - - - - - - - - - - - - - - - - - -

Solution:

Answer:  $x = -2$

- - - - - - - - - - - - - - - - - - - - - - - - - - - - - - - - - - - - - - - - - - - - -

I assume that by now you have learned the basic strategy for solving first-degree equations, and so I have omitted writing out the first four steps in the following examples. However, I will continue to remind you of step 5, that is, to check your tentative solution. Here are two more examples.

## EXAMPLE 3

Solve for $x$: $\dfrac{2(x+1)}{x} = \dfrac{2}{x}$.

Solution:

$$\dfrac{2(x+1)}{x} = \dfrac{2}{x}$$

$$\left(\dfrac{2x+2}{x} = \dfrac{2}{x}\right)x$$

$$\left(\dfrac{2x+2}{x}\right)x = \dfrac{2}{x} \cdot x$$

$$2x + 2 = 2$$

$$2x = 2 - 2$$

$$\dfrac{2x}{2} = \dfrac{0}{2}$$

$$x = 0$$

Check:

$$\dfrac{2(x+1)}{x} = \dfrac{2}{x}$$

$$\dfrac{2(0+1)}{0} \stackrel{?}{=} \dfrac{2}{0}$$

However, recall that we said that **division by zero is undefined**! Therefore $x = 0$ cannot be a solution to this equation.

Answer: The equation has no solution.

What Example 3 should impress on you is the necessity for checking the tentative solution.

## EXAMPLE 4

Solve for $x$: $\dfrac{2x-4}{x-3} = 3 + \dfrac{2}{x-3}$.

Solution:

$$\frac{2x-4}{x-3} = 3 + \frac{2}{x-3}$$

$$\frac{2x-4}{x-3}(x-3) = 3(x-3) + \frac{2}{x-3}(x-3)$$

$$2x-4 = 3(x-3) + 2$$

$$2x-4 = 3x - 9 + 2$$

$$2x-4 = 3x - 7$$

$$2x - 3x = -7 + 4$$

$$-x = -3$$

$$x = 3$$

Check:
$$\frac{2x-4}{x-3} = 3 + \frac{2}{x-3}$$

$$\frac{6-4}{3-3} \stackrel{?}{=} 3 + \frac{2}{3-3}$$

But division by zero is undefined, so 3 cannot be a solution.

Answer:   The equation has no solution.

_ _ _ _ _ _ _ _ _ _ _ _ _ _ _ _ _ _ _ _ _ _ _ _ _ _ _ _ _ _ _ _ _ _ _ _ _

It's your turn again.

## Problem 2

Solve for $x$:   $\dfrac{3}{2x-1} = 5.$

_ _ _ _ _ _ _ _ _ _ _ _ _ _ _ _ _ _ _ _ _ _ _ _ _ _ _ _ _ _ _ _ _ _ _ _ _

Solution:

Answer:   $x = \dfrac{4}{5}$

_ _ _ _ _ _ _ _ _ _ _ _ _ _ _ _ _ _ _ _ _ _ _ _ _ _ _ _ _ _ _ _ _ _ _ _ _

Fractional equations of the following type occur quite often:

$$\frac{2}{x+1} = \frac{3}{x}$$

By this I mean an equation with **two** fractions, one on each side of the equal sign, and no other terms. There is an easy way to solve this kind of fractional equation, that is, to "cross-multiply."

$$\frac{2}{x+1} \diagdown \diagup \frac{3}{x}$$

$$2x = 3(x+1)$$

What we actually did was to multiply the entire equation by the lowest common denominator, $x(x+1)$. But by simply "cross-multiplying" we save ourselves a few steps. (If you don't believe me, try it the long way.) Then the problem continues on as before.

$$2x = 3x + 3$$

$$2x - 3x = 3$$

$$-x = 3$$

$$x = -3$$

Check:

$$\frac{2}{x+1} = \frac{3}{x}$$

$$\frac{2}{(-3)+1} \overset{?}{=} \frac{3}{(-3)}$$

$$\frac{2}{-2} \overset{?}{=} \frac{3}{-3}$$

$$-1 = -1$$

Answer:   This is true. So $x = -3$ is the solution.

_____

===============================================================

## EXAMPLE 5

Solve for $x$:   $\dfrac{5}{2x-1} = \dfrac{3}{x+1}$.

_____

Solution:

$$\frac{5}{2x-1} \diagdown \diagup \frac{3}{x+1}$$

$$5(x+1) = 3(2x-1)$$

$$5x + 5 = 6x - 3$$

$$5x - 6x = -3 - 5$$

$$-x = -8$$

$$x = 8$$

Check:

$$\frac{5}{2x-1} = \frac{3}{x+1}$$

$$\frac{5}{2(8)-1} \overset{?}{=} \frac{3}{(8)+1}$$

$$\frac{1}{3} = \frac{1}{3}$$

Answer:   This is true. So $x = 8$ is the solution.

_____

## EXAMPLE 6

Solve for $x$:  $\dfrac{3x}{2x-3} = 4.$

- - - - - - - - - - - - - - - - - - - - - - - - - - - - - - - - - - - - - - - - - - - -

Solution:  Since a whole number can always be divided by 1 without changing its value, we can write 4 as $\dfrac{4}{1}$ and then cross-multiply.

$$\frac{3x}{2x-3} = \frac{4}{1}$$

$$3x(1) = 4(2x-3)$$

What would have happened if we chose to multiply through by the lowest common denominator instead? Finish the problem yourself.

Answer:  $x = \dfrac{12}{5}$

- - - - - - - - - - - - - - - - - - - - - - - - - - - - - - - - - - - - - - - - - - - -

You should now be able to identify and solve fractional equations. Remember that an equation of this type has a variable in a denominator. In fact, there may be a fraction on either side or both sides of the equal sign.

The same four basic steps—simplify, transpose, simplify, and divide—are used to solve fractional equations. In addition, we must check the tentative solution by substituting it in the original equation.

Remember also that, if we make a wrong choice for the common denominator and a denominator remains after completing the basic steps, we can simply repeat the process, using the remaining denominator.

Finally, remember that, when there is **a single** fraction on both sides of the equal sign, we can use a shortcut: cross-multiplying the fractions.

Before beginning the next unit you should solve the following equations.

## EXERCISES

Solve for $x$:

1.  $1 = \dfrac{5}{x}$

2.  $\dfrac{x-3}{2} = \dfrac{2x+4}{5}$

3.  $\dfrac{6}{x-2} = -3$

4. $\dfrac{3x-3}{x-1} = 2$

5. $\dfrac{x}{2} = \dfrac{x+6}{3}$

6. $\dfrac{3}{x} = \dfrac{4}{x-2}$

7. $\dfrac{4}{x+3} = \dfrac{1}{x-3}$

8. $\dfrac{5-2x}{x-1} = -2$

9. $\dfrac{x+3}{x-2} = 2$

10. $\dfrac{4}{5}x - \dfrac{1}{4} = -\dfrac{3}{2}x$

11. $5 + \dfrac{3+x}{x} = \dfrac{5}{x}$

12. $\dfrac{4}{x-2} - \dfrac{1}{x} = \dfrac{5}{x-2}$

Hint: The common denominator is $x(x-2)$.

# UNIT 5

## Another Special Type of Equation: The Literal Equation

In this unit you will learn about another special type of equation, the literal equation. When you have completed the unit, you will be able to identify and solve literal equations.

> Definition:   A **literal equation** is an equation that contains letters and numbers. **Formulas** often are written as literal equations.

We solve a literal equation for the stated letter in terms of the other letters. That is, our answer will no longer be a simple numerical value but will contain some, or all, of the other letters in the equation.

Again, the basic strategy we have been using to solve other first-degree equations can be used to solve literal equations. I have expanded the steps slightly because there is more than one letter in the equation.

To solve a literal equation for a given letter, identify the letter and

1. **Simplify:**   remove parentheses,
    clear any fractions,
    collect like terms.

2. **Transpose:**
    Isolate all terms with the letter to be solved for on one side of the equation and transpose everything else to the other side.

3. **Simplify.**

4. **Divide** each term of the equation **by the coefficient** of the letter to be solved for.

This is an example of a literal equation:

$$2x - 4p = 3x + 2p$$

Although literal equations may appear at first glance to be more difficult, you will find that they are easier to solve than many other equations you have done.

## EXAMPLE 1

Solve this literal equation for $x$:   $2x - 4p = 3x + 2p$.

Solution:   $2x - 4p = 3x + 2p$

$$2x - 3x = 2p + 4p$$

$$-x = 6p$$

$$\frac{-x}{-1} = \frac{6p}{-1}$$

$$x = -6p$$

Now *you* solve the same equation for $p$.

## Problem 1

Solve this literal equation for $p$:   $2x - 4p = 3x + 2p$.

Solution:

Answer:   $p = \dfrac{-x}{6}$

Now we briefly turn our attention to some examples with formulas. The area of a triangle is given by the formula: $A = \dfrac{1}{2}bh$, where $A$ is the area, $b$ is the base, and $h$ is the altitude.

## EXAMPLE 2

Solve this formula for $b$:    $A = \dfrac{1}{2}bh$.

Solution:    $A = \dfrac{1}{2}bh$

$2\left( A = \dfrac{1}{2}bh \right)$

$2 \cdot A = 2 \cdot \dfrac{1}{2}bh$

$2A = bh$

$\dfrac{2A}{h} = \dfrac{b\cancel{h}}{\cancel{h}}$

$\dfrac{2A}{h} = b$

## Problem 2

This formula is for the perimeter of a rectangle: $P = 2l + 2w$, where $P$ is the perimeter, $l$ is the length, and $w$ is the width.

Solve for $w$.

Solution:

Answer:    $w = \dfrac{P - 2l}{2}$

Let's do two more examples.

## EXAMPLE 3

Solve this equation for $x$:    $ax - 3 = -2cx$

Solution:                $ax - 3 = -2cx$

$ax - 3 = -2cx$

Transpose:    $ax + 2cx = 3$

Recall from Unit 2 the distributive property, $a(b + c) = ab + ac$, which we used to remove parentheses. Now, in order to find the coefficient of $x$, we will use the distributive property in reverse.

Notice that $x$ is common to both terms on the left side of the equation. Using the distributive property, we can factor out the common $x$ since

$$ax + 2cx = x(a + 2c)$$

Factor out $x$:　$x(a + 2c) = 3$

$$\frac{x(\cancel{a + 2c})}{\cancel{a + 2c}} = \frac{3}{a + 2c}$$

$$x = \frac{3}{a + 2c}$$

Notice that for our literal equations we do *not* get a nice simple number for an answer; instead we get a fairly complicated-looking expression for $x$ in terms of the other letters.

----

# EXAMPLE 4

Solve this equation for $x$:　$3x + 5y = ax + 2y$.

----

Solution:

$$3x + 5y = ax + 2y$$
$$3x - ax = 2y - 5y$$
$$3x - ax = -3y$$

Factor out $x$.

$$x(3 - a) = -3y$$

$$\frac{x(\cancel{3 - a})}{\cancel{3 - a}} = \frac{-3y}{3 - a}$$

$$x = \frac{-3y}{3 - a}$$

Or, if you prefer, multiply top and bottom by $-1$.

$$x = \frac{3y}{a - 3}$$

----

More often than not we solve equations for $y$ in terms of $x$. Using the same equation as in Example 4, try it.

---

## Problem 3

Solve this equation for $y$: $3x + 5y = ax + 2y$.

- - - - - - - - - - - - - - - - - - - - - - - - - - - - - - - - -

Solution:

Answer: $y = \dfrac{ax - 3x}{3}$

- - - - - - - - - - - - - - - - - - - - - - - - - - - - - - - - -

We'll do two more examples.

====================================================

## EXAMPLE 5

Solve this equation for $y$: $\dfrac{3y + a}{a} = \dfrac{4y + b}{b}$.

- - - - - - - - - - - - - - - - - - - - - - - - - - - - - - - - -

Solution:
$$\frac{3y + a}{a} = \frac{4y + b}{b}$$

Shortcut: cross-multiply from Unit 4.    $b(3y + a) = a(4y + b)$

$$3by + ab = 4ay + ab$$

$$3by - 4ay = ab - ab$$

Factor out $y$.    $y(3b - 4a) = 0$

Divide by the coefficient of $y$.    $\dfrac{y(3b - 4a)}{(3b - 4a)} = \dfrac{0}{3b - 4a}$

$$y = 0$$

QUESTION: Must we now check to see whether this answer is indeed the solution to the equation? In other words, is this a fractional equation?

- - - - - - - - - - - - - - - - - - - - - - - - - - - - - - - - -

## EXAMPLE 6

Solve for $x$:   $a(x-1) = -\dfrac{x}{b}$.

Solution:

$$a(x-1) = -\frac{x}{b}$$

$$ax - a = -\frac{x}{b}$$

$$\left(ax - a = \frac{-x}{b}\right)b$$

$$abx - ab = -x$$

$$abx + x = ab$$

Factor out $x$:   $x(ab+1) = ab$

$$\frac{x(ab+1)}{(ab+1)} = \frac{ab}{ab+1}$$

$$x = \frac{ab}{ab+1}$$

Now it's your turn to solve the above equation for $b$.

## Problem 4

Solve for $b$:   $a(x-1) = -\dfrac{x}{b}$.

Solution:

Answer:   $b = \dfrac{-x}{a(x-1)}$

You should now be able to identify and solve first-degree literal equations for a particular letter. Remember that a literal equation is an equation that contains numbers and more than one letter. Formulas are often written as literal equations. Even though literal equations are a special type of equation, the same basic strategy is used to solve them.

Remember also that, as we learned in Unit 4, when there is a **single** fraction on both sides of a literal equation, we can reduce our work by cross-multiplying.

Before beginning the next unit you should solve the following literal equations.

## EXERCISES

Solve for $a$:

1. $\dfrac{2ax}{3c} = \dfrac{y}{m}$

2. $2cy + 4d = 3ax - 4b$

3. $ax + 3a = bx + 7c$

4. $4x + 5c - 2a = 0$

5. $a(x + 2) = \pi - cy$

6–10. Now solve each of the above equations for $x$.

11–15. Now solve each of the above equations for $c$.

16. $C = 2\pi r$ is the formula for the circumference of a circle. Solve for $r$.

17. $R = \dfrac{V}{I}$ is Ohm's law in electrical theory. Solve for $I$.

18. $A = \dfrac{1}{2}(a + b)h$ is the formula for the area of a trapezoid. Solve for $a$.

# UNIT 6

## Applied Problems

In this unit we will consider a few applied problems, sometimes referred to as word problems. First, you will learn how to translate a word problem into an algebraic equation, which then can be solved using the same basic strategy learned in previous units. The emphasis will be not only on the numerical answer but also on the interpretation of the answer in terms of the problem.

If you're tempted to skip this unit, don't. The problems really aren't that hard, and it is important to understand how algebra can be used to solve problems in the real world. I like to think of each one as a puzzle, and I enjoy solving puzzles.

## CHANGING VERBAL STATEMENTS TO ALGEBRAIC EXPRESSIONS

In algebra, as you know, letters are used to represent numbers. By using letters and mathematical symbols, we can replace lengthy verbal statements with short algebraic expressions.

Here are a few examples.

| EXAMPLE 1 | The sum of a number and 7 | $x + 7$ |
| EXAMPLE 2 | Five minus some number | $5 - y$ |
| EXAMPLE 3 | Seven times a number plus 3 | $7x + 3$ |
| EXAMPLE 4 | A number divided by 11 | $\dfrac{z}{11}$ |

You try the next few.

Let $x$ represent a number. Write each of the following in terms of $x$.

Problem 1    Three times a number minus 2

Problem 2    Seven divided by a number

Problem 3    Five plus a number

Answers:   1. $3x - 2$    2. $\dfrac{7}{x}$    3. $5 + x$

# CHANGING VERBAL STATEMENTS TO ALGEBRAIC EQUATIONS

We'll continue translating words into symbols, but now we'll include an equal sign.

EXAMPLE 5    Three times a number minus 2 equals 10.

Solution:   $3x - 2 = 10$

EXAMPLE 6    Five times a number plus 2 is 10.

Solution:   $5x + 2 = 10$

Notice that the word *is* translates to an equal sign.

Here are three for you to try. For each problem, write an equation and then solve it.

Problem 4    Five plus a number is 7.

Answer:  $5 + x = 7$,  $x = 2$

----

Problem 5    When a number is decreased by 3, the result is 15.

Answer:  $x - 3 = 15$,  $x = 18$

----

Problem 6    The sum of a number and 11 is 12.

Answer:  $x + 11 = 12$,  $x = 1$

----

I think we're ready to look at some word problems.

# SOLVING WORD PROBLEMS

When attempting to solve a word problem, we must translate the problem into algebraic symbols, write and solve the equation, and finally answer the original question. I'll demonstrate how to solve one or two word problems before I give you a similar problem to solve.

## EXAMPLE 7

Bob is 5 years older than Barbara. The sum of their ages is 23. How old is Barbara?

---

Solution:   Let $x$ represent Barbara's age.

Then Bob's age is $x + 5$ because he is 5 years older than Barbara.

The sum of their ages is 23.

$$\text{Bob's age} + \text{Barbara's age} = 23$$
$$(x + 5) + \qquad\quad x = 23$$
$$2x + 5 = 23$$
$$2x = 18$$
$$x = 9$$

Answer:   Barbara is 9 years old.

---

Maybe you wanted to start the previous example by letting $x$ represent Bob's age. Would this be wrong? No, because applied problems can often be approached from different directions, all of which lead to the correct solution. The next example should help convince you of that fact.

## EXAMPLE 8

Resolve Example 7 letting $x$ represent Bob's age.

---

Solution:   Let $x$ represent Bob's age.

Then Barbara's age is $x - 5$ because she is 5 years younger than Bob.

The sum of their ages is 23.

$$\text{Bob's age} + \text{Barbara's age} = 23$$
$$x + (x - 5) = 23$$
$$2x - 5 = 23$$
$$2x = 28$$
$$x = 14$$

But the problem asked for Barbara's age.

Barbara's age is given by $x - 5$, thus $14 - 5 = 9$.

Answer:   Barbara is 9 years old.

## Problem 7

Laurie and Lynda are my husband's two daughters. Lynda is 16 years older than Laurie. The sum of their ages is 60. How old is each daughter?

Answer: Laurie is 22 years old and Lynda is 38 years old.

Before continuing we need to introduce some new terminology with respect to integers. Recall that the set of integers = {. . . , –4, –3, –2, –1, 0, 1, 2, 3, 4, . . .}.

Examples of **consecutive integers** are 21, 22, 23, 24 and –5, –4, –3. Notice that if $x$ represents an integer, a set of consecutive integers can be represented as $x$, $x + 1$, $x + 2$, $x + 3$, and so on.

Examples of **consecutive even integers** are 10, 12, 14, 16 and –4, –2, 0, 2, 4. Notice that if $x$ represents an even integer, a set of even consecutive integers can be represented as $x$, $x + 2$, $x + 4$, and so on.

Examples of **consecutive odd integers** are 37, 39, 41 and –7, –5, –3, –1, 1, 3. Notice that if $x$ represents an odd integer, a set of consecutive odd integers can be represened as $x$, $x + 2$, $x + 4$, and so on. Consecutive odd integers, like consecutive even integers, always differ by 2.

## EXAMPLE 9

On U.S. Route 1 in the Floride Keys, the mile marker numbers increase from the south (Key West) to the north (Key Largo). The sum of two consecutive mile markers on U.S. Route 1 in the Florida Keys is 115. Find the numbers on the markers.

Solution: Let $x$ represent the smaller number.

Then the next consecutive integer is $x + 1$.

The sum of the two consecutive mile markers is 115.

$$x + (x + 1) = 115$$
$$2x + 1 = 115$$
$$2x = 114$$
$$x = 57$$
$$\text{and } x + 1 = 58$$

Answer: The mile marker numbers are 57 and 58.

## EXAMPLE 10

Jerry and Caroline Cash ordered tickets for the Paradise Big Band's final concert of the season. The San Carlos Institute, a small auditorium where concerts are held, has one center aisle. Seats on the right of the aisle are numbered by consecutive even numbers, and on the left by consecutive odd numbers. When the tickets arrive, Jerry notices that the sum of the two seat numbers in 208. Are the Cashes' seats on the right or left of the center aisle and what are their seat numbers?

---------------------------------------------------------------

Solution:   Let $x$ represent the first seat number.

Then the next seat number is $x + 2$ because it is either a consecutive even or consecutive odd number.

The sum of the two seat numbers is 208.

$$x + (x + 2) = 208$$
$$2x + 2 = 208$$
$$2x = 206$$
$$x = 103$$
$$x + 2 = 105$$

Answer:   Jerry and Caroline Cash will be seated on the left side of the aisle in seats 103 and 105.

---------------------------------------------------------------

I have done two short examples; now here's a longer one for you.

## Problem 8

Rob, my husband's youngest son and my favorite, lives in Pittsburgh. His baseball team is raffling off a ten-speed bike to raise money to update the team's mascot. Rob bought five raffle tickets, which are numbered consecutively. The sum of the raffle ticket numbers is 1075. What are the numbers on his raffle tickets?

---------------------------------------------------------------

Solution:

Answer:   The numbers on the raffle tickets are 213, 214, 215, 216, and 217.

---------------------------------------------------------------

Many application problems involve formulas we already know. A familiar formula that relates the concepts of distance, speed, and time is $rt = d$. For example, if you drive on the turnpike for 2 hours at an average speed of 60 mph, you will have traveled a distance of 120 miles, or 2 times $60 = 120$. We will use this formula in the next two examples, which are often referred to as motion problems.

## EXAMPLE 11

Patti and Tom decide to go for a walk. Patti likes to walk fast, and Tom likes to walk slow. Patti starts out walking due north at 3 mph, while Tom decides to head due south, ambling along at 1 mph. After how many hours will they be 5 miles apart?

Solution:  I find it helpful to construct a chart for motion problems of this type. The formula is shown across the top, and the given information about the rates has been filled in.

|       | rate × | time | = distance |
|-------|--------|------|------------|
| Patti | 3 mph  |      |            |
| Tom   | 1 mph  |      |            |

Let $t$ represent Patti's walking time in hours.

Then Tom's walking time also will be $t$ because they are starting and stopping at the same time.

The final column of the chart is found by multiplying across each row, that is, rate × time = distance.

|       | rate × | time | = distance |
|-------|--------|------|------------|
| Patti | 3 mph  | $t$  | $3t$       |
| Tom   | 1 mph  | $t$  | $t$        |

Now we are ready to try writing the equation needed to solve the problem. Typically the required information comes from the two entries in the distance column. You need to ask yourself a few questions. Will the distances walked by Patti and Tom be equal? No, because they are walking at different rates but for the same amount of time. Do we know what the sum of the distances walked will equal? Yes, the sum of the distances walked will be 5 miles because we want to know when they will be 5 miles apart.

$$\text{Patti's distance} + \text{Tom's distance} = 5$$
$$3t + t = 5$$
$$4t = 5$$
$$t = \frac{5}{4}$$
$$t = 1.25$$

Answer:     After walking for 1.25 hours (or 1 hour and 15 minutes), Patti and Tom will be 5 miles apart.

_____

The final example, while still using $d = rt$, has a few twists to it, so be careful.

## EXAMPLE 12

Orvis Kemp owns a 1968 red Mercedes roadster in mint condition. Orvis leaves his home in Key West and drives at a constant rate of 40 mph. One hour later, Rita, his wife, leaves in her new Mercedes and travels on the same route at a constant rate of 50 mph. How long will it take for Rita to catch up with Orvis?

_____

Solution:     We're going to work this one together. Without looking at the completed chart, try filling in the blank chart on your own.

|       | rate × | time | = distance |
|-------|--------|------|------------|
| Orvis |        |      |            |
| Rita  |        |      |            |

|       | rate × | time  | = distance |
|-------|--------|-------|------------|
| Orvis | 40 mph | $t$   | $40t$      |
| Rita  | 50 mph | $t - 1$ | $50(t - 1)$ |

Did you realize that in this example the times are not equal? Rita starts out 1 hour later; thus she will be driving 1 hour less than Orvis. Now for the equation needed to solve the problem. What is true about the distances driven by each person? In this problem the distances are equal.

distance driven by Orvis = distance driven by Rita

$$40t = 50(t - 1)$$
$$40t = 50t - 50$$
$$50 = 50t - 40t$$
$$50 = 10t$$
$$\frac{50}{10} = \frac{10t}{10}$$
$$5 = t$$

Answer:   Rita will catch up with Orvis in 4 hours. Note that the answer is 4, not 5, because the question asked about Rita's driving time, which was $t - 1$.

---

I think we've done enough examples.

By now you should be able to translate verbal expressions into algebraic symbols, and to translate an applied problem into an equation.

Before going on to the next unit, do the following exercises.

## EXERCISES

Let $x$ represent a number. Express each of the following in terms of $x$:

1.   A number decreased by 5.

2.   Three times a number increased by 8.

3.   Eight times a number minus 10.

4.   A number divided by 3.

Let $x$ represent a number. Express each of the following as an equation:

5.   Two times a number decreased by 5 equals 11.

6.   Seven times a number is 35.

7.   A number added to 20 is the same as 32.

8.   The sum of $x$ and 12 is 20.

9.   Fifteen increased by 2 times a number is 47.

10.   Four more than three times a number is 17.

Solve each of the following:

11.   George is 8 years older than Jack. The sum of their ages is 42. How old is each person?

12.   A rope that is 36 feet in length is cut into two pieces. If one piece is 10 feet longer than the other, how long is each piece?

13. The Cancer Foundation of Denver is raffling off a PT Cruiser. The raffle tickets are $100 each and are tax deductible. Mercy Hiller and her husband have bought four tickets, which they notice are numbered consecutively. The sum of the numbers is 1354. What are the numbers on their raffle tickets?

14. The sum of three consecutive even integers is 6480. What are the integers?

15. The formula for the perimeter of a rectangle is $P = 2l + 2w$. The perimeter of the local high school basketball court is 268 ft. The length is 34 ft longer than the width. Find the dimensions of the basketball court.

16. The perimeter of the Searstown parking lot is 4100 ft. The length is 3 times the width. Find the dimensions of the parking lot.

17. Aurora and Chris work together at the local campus bookstore. They live 4 miles apart on the same road. Aurora starts walking at 3 mph, and at the same instant Chris starts walking toward her at 2.2 mph. When will they meet?

18. Every morning Rosalie runs along a runner's path in Pawley's Island at the constant rate of 6 mph. A half-hour later her friend, Sally, begins at the same point, running at a rate of 8 mph and following the same path. How long will it take Sally to reach Rosalie?

# UNIT 7

## Positive Integral Exponents

In this unit you will learn to simplify expressions involving positive integral exponents. You will also learn four of the **five basic laws of exponents**. When you have completed the unit, you will be able to simplify expressions in which there is an exponent that is a positive integer or 0.

## RECOGNIZING EXPONENTS

Consider the expression $b^n$. This is read as "$b$ to the $n$th power." The $b$ is referred to as the **base**, and the $n$ is the **exponent**. Here $b \neq 0$.

In this unit we will consider only exponents that are 0 or a positive integer (1, 2, 3, 4, etc.). Negative and fractional exponents will be discussed in later units.

---

Definition:   **Positive Integral Exponent**

$$b^n = \underbrace{b \cdot b \cdot b \cdot \ldots \cdot b}_{n \text{ factors}}, \text{ if } n \text{ is a positive integer}$$

---

Consider the positive integral exponents in these expressions.

---

EXAMPLE 1  $b^2 = b \cdot b$

EXAMPLE 2  $x^3 = x \cdot x \cdot x$

EXAMPLE 3  $2^5 = 2 \cdot 2 \cdot 2 \cdot 2 \cdot 2 = 32$

- - - - - - - - - - - - - - - - - - - - - - - - - - - - - - - - -

Any nonzero quantity raised to the 0 power is 1.

> Definition:  **Zero Exponent**
>
> $b^0 = 1$, where $b \neq 0$

---

EXAMPLE 4  $x^0 = 1$

EXAMPLE 5  $5^0 = 1$

EXAMPLE 6  $\left(3ab + \pi - 5\sqrt{7}\right)^0 = 1$

- - - - - - - - - - - - - - - - - - - - - - - - - - - - - - - - -

# SIMPLIFYING EXPRESSIONS WITH EXPONENTS

To accomplish simplification, we have five basic laws of exponents. We will discuss only four at this time.

These laws are used to shorten our work. When in doubt about any of these laws, we can always go back to the definitions on the first and second pages of this unit.

> **Laws of Exponents**
>
> I. **Multiplication**  $b^n \cdot b^m = b^{n+m}$
>
> II. **Power of a power**  $(b^n)^m = b^{nm}$
>
> III. **Power of a product**  $(bc)^n = b^n c^n$
>
> IV. **Power of a fraction**  $\left(\dfrac{a}{b}\right)^n = \dfrac{a^n}{b^n}$

> I. **Multiplication**  $b^n \cdot b^m = b^{n+m}$

## EXAMPLE 7

Simplify: $a^2 \cdot a^3$.

---

Solution: $a^2 \cdot a^3 = a^{2+3} = a^5$

If you doubt this, use the definition: $a^2 \cdot a^3 = (a \cdot a) \cdot (a \cdot a \cdot a) = a^5$.

---

Note that, as in preceding units, if no number is written, in this case an exponent, the number is understood to be 1.

## EXAMPLE 8

Simplify: $x \cdot x^2$.

---

Solution: $x \cdot x^2 = x^1 \cdot x^2 = x^{1+2} = x^3$

---

## EXAMPLE 9

Simplify: $5 \cdot 5^2$.

---

Solution: $5 \cdot 5^2 = 5^1 \cdot 5^2 = 5^{1+2} = 5^3$

In case you're not convinced:

$$5 \cdot 5^2 = 5 \cdot (5 \cdot 5) = 5^3$$

---

## EXAMPLE 10

Simplify: $a^2(2a^3)$.

---

Solution: $a^2(2a^3) = a^2 \cdot 2 \cdot a^3 = 2a^5$

---

Before doing the next example, here is a reminder from Unit 2. When a term contains a number and several letters, the number is written first and the letters usually are put in alphabetical order.

## EXAMPLE 11

Simplify: $(a^2b^3)(ab^4)(3abc)$.

---

Solution: $(a^2b^3)(ab^4)(3abc)$

$a^2 \cdot b^3 \cdot a^1 \cdot b^4 \cdot 3 \cdot a^1 \cdot b^1 \cdot c^1 = 3a^{2+1+1}b^{3+4+1}c^1 = 3a^4b^8c$

---

You try some.

---

Problem 1    $x^3 \cdot x^5 =$

Problem 2    $x^2(x^3 y) =$

Problem 3    $(x^2 y^3)(x^7 y) =$

Problem 4    $(2w^3 a^5)(3a^2 w)$

Answers:  1. $x^8$    2. $x^5 y$    3. $x^9 y^4$    4. $6a^7 w^4$

---

> II.  **Power of a power**    $(b^n)^m = b^{nm}$

## EXAMPLE 12

Simplify:  $(a^2)^3$.

---

Solution:  $(a^2)^3 = a^{2 \cdot 3} = a^6$

If we had used the definition instead:

$$(a^2)^3 = (a^2)(a^2)(a^2)$$
$$= a^{2+2+2}$$
$$= a^6$$

---

Note that the laws of exponents are shortcut methods. With them, you do not have to work out expressions completely with the definition. But they are shortcuts only if you apply them properly.

## EXAMPLE 13

Simplify:  $(x^{15})^2$.

---

Solution:  $(x^{15})^2 = x^{15 \cdot 2} = x^{30}$

---

Now it's your turn again.

---

Problem 5    $(a^{10})^2 =$

Problem 6    $(x^3)^0 =$

Problem 7    $c^{10} \cdot c^2 =$

Problem 8    $(w^2)^4$

Answers:  5. $a^{20}$    6. 1    7. $c^{12}$    8. $w^8$

> III. **Power of a product**   $(bc)^n = b^n c^n$

Notice that this law is really just an extension of Law II: $(b^n)^m = b^{nm}$.

## EXAMPLE 14

Simplify:   $(3x^2)^2$.

---

Solution:   $(3x^2)^2 = 3^2 x^4 = 9x^4$

Be careful; most people forget that the 3 must also be squared.

---

## EXAMPLE 15

Simplify:   $(2x^2y^3)^4$.

---

Solution:   $(2x^2y^3)^4 = 2^4 x^{2 \cdot 4} y^{3 \cdot 4} = 2^4 x^8 y^{12} = 16x^8 y^{12}$

---

> IV. **Power of a fraction**   $\left(\dfrac{a}{b}\right)^n = \dfrac{a^n}{b^n}$

Notice that this law too is really just an extension of Law II: $(b^n)^m = b^{nm}$.

## EXAMPLE 16

Simplify:   $\left(\dfrac{x^2}{c^3}\right)^5$.

---

Solution:   $\left(\dfrac{x^2}{c^3}\right)^5 = \dfrac{x^{2 \cdot 5}}{c^{3 \cdot 5}} = \dfrac{x^{10}}{c^{15}}$

---

## EXAMPLE 17

Simplify:   $\left(\dfrac{7w^3}{5s}\right)^2$.

---

Solution:   $\left(\dfrac{7w^3}{5s}\right)^2 = \dfrac{7^2 w^6}{5^2 s^2} = \dfrac{49w^6}{25s^2}$

---

Have you noticed that Laws II, III, and IV deal with removing parentheses? To remove the parentheses we multiply the exponents.

Sound familiar? To remove parentheses, we multiply. . . .

As the algebraic expressions become more complex, and you know they will, it often becomes necessary to use several laws in the same problem. My choice of procedure **always** will be to remove the parentheses first (Laws II, III, and IV) and finish simplifying the expression by using Law I.

## EXAMPLE 18

Simplify: $a^2(4a^3)^2$.

---

Solution: $a^2(4a^3)^2 = a^2 \cdot 4^2a^6 = 16 \cdot a^{2+6} = 16a^8$

As in solving equations, I removed the parentheses first.

---

## EXAMPLE 19

Simplify: $x^3(3\pi x)^0$.

---

Solution: $x^3(3\pi x)^0 = x^3 \cdot 1 = x^3$

---

Try another yourself.

## Problem 8

Simplify: $x^3(xy^2)^2$.

---

Solution:

Answer: $x^5y^4$

---

You should now be able to simplify any expression in which an exponent of 0 or some positive integer appears. Remember that, when you are in doubt about one of the laws of exponents, you can always return to the definition of a positive integral exponent or a zero exponent.

Now try to simplify the following expressions. Be sure to check your answers with those at the back of the book and correct any mistakes before continuing on to the next unit.

## EXERCISES

Simplify:

1.  $(3y)^2$

2.  $3x^0$

3.  $x^2(x^3)^4$

4.  $(x^2y^3z)^4$

5.  $\left(\dfrac{x^2}{wz}\right)^3$

6.  $(2ab)b^2$

7.  $\dfrac{(x^2y^3)^2}{5}$

8.  $5(x^2z)^2$

9.  $(5x^2z)^2$  Be sure to notice that Problems 8 and 9 are different.

10.  $\left(\dfrac{a^2b^3cd^5}{3x^2w^0}\right)^7$

11.  $\dfrac{(2ab)^2}{(3x^3)^2}$

12.  $(3x^5)^2(2x^3)^3$

13.  $(x^2y)(xy^2)$

14.  $2(3ab^2)^2$

15.  $(-4c)^2$

16.  $\left(\dfrac{xyz^2}{5a}\right)^3$

17.  $(-2abc)(bcd)(3abc^2)$

18.  $(2x^2yz)(-5xz)^2(xyz^2)^3$

# UNIT 8

## Negative Exponents

In this unit you will learn how to work with **negative exponents**. When you have completed the unit, you will be able to simplify expressions with negative exponents.

> Definition:   **Negative exponent**
>
> $$b^{-n} = \frac{1}{b^n} \quad \text{and} \quad \frac{1}{b^{-n}} = b^n$$
>
> where $b \neq 0$

The right-hand part of the above statement follows from the fact that

$$\frac{1}{b^{-n}} = \frac{1}{\dfrac{1}{b^n}} = 1 \div \frac{1}{b^n} = 1 \cdot b^n = b^n$$

Therefore a **factor** with a negative exponent can be rewritten with a positive exponent by moving the **factor** from the numerator to the denominator (or vice versa).

Consider Examples 1–12 and be certain you understand the simplification.

EXAMPLE 1   $2^{-1} = \dfrac{1}{2}$

EXAMPLE 2   $a^{-1} = \dfrac{1}{a}$

EXAMPLE 3   $b^{-2} = \dfrac{1}{b^2}$

EXAMPLE 4   $c^{-3} = \dfrac{1}{c^3}$

EXAMPLE 5   $2x^{-1} = \dfrac{2}{x}$

EXAMPLE 6   $3ab^{-2} = \dfrac{3a}{b^2}$

EXAMPLE 7   $5^{-1}ab^{-2}c = \dfrac{ac}{5b^2}$

EXAMPLE 10   $\dfrac{a^{-2}}{b^{-3}} = \dfrac{b^3}{a^2}$

EXAMPLE 8   $\dfrac{1}{a^{-2}} = a^2$

EXAMPLE 11   $\dfrac{y^{-3}}{x^2} = \dfrac{1}{y^3 x^2}$

EXAMPLE 9   $\dfrac{5}{x^{-3}y} = \dfrac{5x^3}{y}$

EXAMPLE 12   $\dfrac{-3x^2 z}{w} = \dfrac{-3z}{wx^2}$

A common mistake is to think that the −3 in Example 12 can be rewritten as a positive 3 by moving it to the denominator of the fraction. This would be wrong! The −3 is not an exponent; consequently the laws of exponents do not apply.

––––––––––––––––––––––––––––––––––––––––––––––––––––

When simplifying expressions dealing with exponents, the objective is to write the final answer *without* zero or negative exponents. Luckily the laws of exponents introduced in Unit 7 apply for all types of exponents. We can use these laws and our definition to simplify expressions with negative exponents.

Laws of Exponents

I.   **Multiplication**        $b^n \cdot b^m = b^{n+m}$         add exponents

II.  **Power of a power**     $(b^n)^m = b^{nm}$

III. **Power of a product**  $(bc)^n = b^n c^n$        multiply exponents

IV.  **Power of a fraction**  $\left(\dfrac{a}{b}\right)^n = \dfrac{a^n}{b^n}$

With the introduction of negative exponents, algebraic expressions can become even more complex. Often there are several ways to simplify such complicated expressions. To make it easier for you to follow my solutions, I **always** will use this procedure when simplifying:

1.  If there are parentheses, remove them using Laws II, III, and IV.

2.  If there are negative exponents, rewrite them as positive exponents using the definition.

3.  If necessary, finish by using the multiplication law of exponents, Law I.

Before proceeding you might find it helpful to quickly review Unit 1 on adding, subtracting, dividing, and especially multiplying signed numbers.

## EXAMPLE 13

Simplify:   $(10^{-3})^2$.

––––––––––––––––––––––––––––––––––––––––––––––––––––

Solution:   $(10^{-3})^2 = 10^{-6} = \dfrac{1}{10^6}$

––––––––––––––––––––––––––––––––––––––––––––––––––––

## EXAMPLE 14

Simplify: $(xy^{-1})^{-3}$.

Solution: $(xy^{-1})^{-3} = x^{-3}y^3 = \dfrac{y^3}{x^3}$

## EXAMPLE 15

Simplify: $\dfrac{ab^{-4}}{a^{-2}b}$.

Solution: $\dfrac{ab^{-4}}{a^{-2}b} = \dfrac{a \cdot a^2}{b^4 \cdot b} = \dfrac{a^3}{b^5}$

## EXAMPLE 16

Simplify: $\left(\dfrac{x^2}{y^{-3}w}\right)^{-1}$

Solution: $\left(\dfrac{x^2}{y^{-3}w}\right)^{-1} = \dfrac{x^{-2}}{y^3 w^{-1}} = \dfrac{w}{x^2 y^3}$

Now try four short problems on your own.

## Problem 1

Simplify: $3a^{-2}x^5$.

Solution:

Answer: $\dfrac{3x^5}{a^2}$

## Problem 2

Simplify: $\dfrac{7x^{-3}wz^2}{x^4}$.

Solution:

Answer: $\dfrac{7wz^2}{x^7}$

## Problem 3

Simplify: $\left(\dfrac{3xy^{-1}}{y}\right)^2$.

Solution:

Answer: $\dfrac{9x^2}{y^3}$

## Problem 4

Simplify: $\left(\dfrac{a}{b}\right)^{-1}$.

Solution:

Notice what happened—the fraction is simply inverted. *You have just proved a theorem!*

Hence

$$\left(\frac{a}{b}\right)^{-1} = \frac{b}{a}$$

We'll end the unit with three more examples.

EXAMPLE 17 $\quad \left(\dfrac{3}{4}\right)^{-1} = \dfrac{4}{3}$

EXAMPLE 18 $\quad \left(\dfrac{a^2b}{c^3}\right)^{-1} = \dfrac{c^3}{a^2b}$

Try the next one yourself before looking at the answer.

EXAMPLE 19

Simplify: $\left(\dfrac{x^2y^0}{w^{-1}}\right)^{-2}$.

Solution: $\left(\dfrac{x^2y^0}{w^{-1}}\right)^{-2} = \dfrac{x^{-4}y^0}{w^2} = \dfrac{1y^0}{w^2x^4} = \dfrac{1}{w^2x^4}$

You should now be able to simplify expressions involving negative exponents. Try simplifying the expressions in the exercises by writing them without negative exponents or zero exponents or parentheses.

## EXERCISES

Simplify:

1. $\dfrac{a^{-3}}{a^2}$

2. $\dfrac{a^{-2}x^3}{y^{-1}}$

3. $(x^2y)^{-2}$

4. $x^5 \cdot x^0 \cdot z^{-7}$

5. $\dfrac{x^{-2}}{y^{-3}}$

6. $-2x^6y^0$

7. $(3x^{-6}y^5)^{-2}$

8. $\dfrac{(ab^2)^{-3}}{(x^2y^{-3})^4}$

9. $\dfrac{(3ab^5)^{-3}}{2x^{-5}}$

10. $\dfrac{7x^{-1}}{y^2}$

11. $(5w^{-2})^2(2w^{-2})$

12. $\dfrac{x^{-2}y^{-3}}{(c)^{-2}}$

13. $\dfrac{(5a^2b^3)^2}{(-2x)^{-3}}$

14. $\dfrac{16w^{-1}y^2z^{-3}}{2x}$

15. $\left[\dfrac{b^2}{(a^2b)^{-2}}\right]^{-1}$

# UNIT 9

## Division of Powers

In this unit you will learn to simplify expressions in which the same variables with exponents appear in both the numerator and the denominator. When you have completed the unit, you will be able to simplify expressions involving **division of like variables** raised to integral exponents.

Recall the four laws of exponents from Unit 7.

### Laws of Exponents

I. **Multiplication** $\qquad b^n \cdot b^m = b^{n+m}$ $\qquad$ add exponents

II. **Power of a power** $\quad (b^n)^m = b^{nm}$

III. **Power of a product** $\ (bc)^n = b^n c^n$ $\qquad$ multiply exponents

IV. **Power of a fraction** $\ \left(\dfrac{a}{b}\right)^n = \dfrac{a^n}{b^n}$

We will now add a fifth and final law of exponents, which deals with division:

V. **Division:** $\quad \dfrac{b^m}{b^n} = b^{m-n}$

$\qquad\qquad$ or, alternatively, $\quad$ subtract exponents

$\qquad\qquad = \dfrac{1}{b^{n-m}}$

From a brief analysis of the laws we can produce a convenient way to classify them.

  I.      Deals with multiplication—exponents added.

  II.

  III.          "In a sense" deal with removing parentheses—exponents multiplied.

  IV.

  V.     Deals with division—exponents subtracted.

---

## EXAMPLE 1

Simplify: $\dfrac{x^5}{x^3}$.

---

Solution: $\dfrac{x^5}{x^3} = x^{5-3} = x^2$

An alternative and longer solution uses the definition of positive integral exponents:

$$\frac{x^5}{x^3} = \frac{x \cdot x \cdot \cancel{x} \cdot \cancel{x} \cdot \cancel{x}}{\cancel{x} \cdot \cancel{x} \cdot \cancel{x}} = x \cdot x = x^2$$

---

## EXAMPLE 2

Simplify: $\dfrac{a^4}{a^3}$.

---

Solution: $\dfrac{a^4}{a^3} = a^{4-3} = a^1 = a$

---

What about

$$\frac{c^{15}}{c^3} = ?$$

So far we have been using only the first half of the division law of exponents. Now consider a situation where the exponent in the denominator is larger than the exponent in the numerator.

Recall that, when simplifying an expression, we are attempting to write the final answer without using zero or negative exponents or parentheses.

## EXAMPLE 3

Simplify: $\dfrac{x}{x^4}$.

---

Solution: $\dfrac{x}{x^4} = \dfrac{x^1}{x^4} = x^{1-4} = x^{-3} = \dfrac{1}{x^3}$

or, alternatively,

$$\frac{x}{x^4} = \frac{1}{x^{4-1}} = \frac{1}{x^3}$$

When the larger exponent is in the denominator, it takes one less step to use the second half of the division law of exponents—subtracting in the denominator.

---

## EXAMPLE 4

Simplify: $\dfrac{a^2}{a^{10}}$.

---

Solution: $\dfrac{a^2}{a^{10}} = \dfrac{1}{a^{10-2}} = \dfrac{1}{a^8}$

---

Now consider these examples, which have more than one variable.

## EXAMPLE 5

Simplify: $\dfrac{a^2 b^7 c^3}{a^5 b^2 c^4}$.

---

Solution: $\dfrac{a^2 b^7 c^3}{a^5 b^2 c^4} = \dfrac{b^{7-2}}{a^{5-2} c^{4-3}} = \dfrac{b^5}{a^3 c}$

---

## EXAMPLE 6

Simplify: $\dfrac{a^2 b c^3}{a^7 b^3 c^3}$.

---

Solution: $\dfrac{a^2 b c^3}{a^7 b^3 c^3} = \dfrac{1}{a^{7-2} b^{3-1} c^{3-3}} = \dfrac{1}{a^5 b^2}$

The $c$ does not appear since $c^{3-3} = c^0 = 1$.

---

Again, let's move on to some problems involving definitions and the five laws of exponents.

To accomplish this, as in previous units, I **always** will use the same procedure when simplifying. Of course, an additional step has been added to include division.

My suggested procedure for simplifying an algebraic expression with exponents is:

1.  If there are parentheses, remove them using Laws II, III, and IV.

2.  If there are negative exponents, use the definition to rewrite them as positive exponents.

3.  If applicable, use the division law of exponents, Law V.

4.  If applicable, finish by using the multiplication law of exponents, Law I.

There is nothing sacred about the order of these steps. In fact, they are completely interchangeable. You may do them in any order you wish. I suggest only that you *establish a pattern of your own and stick with it.*

## EXAMPLE 7

Simplify: $\left(\dfrac{xy^0}{x^{-5}}\right)^{-2}$.

Solution: $\dfrac{x^{-2}y^0}{x^{10}} = \dfrac{x^{-2}}{x^{10}} = \dfrac{1}{x^2x^{10}} = \dfrac{1}{x^{12}}$

## EXAMPLE 8

Simplify: $\dfrac{\left(2^2\right)^{-1}}{\left(2^{-4}\right)^2}$.

Solution: $\dfrac{\left(2^2\right)^{-1}}{\left(2^{-4}\right)^2} = \dfrac{2^{-2}}{2^{-8}} = \dfrac{2^8}{2^2} = 2^{8-2} = 2^6$

## EXAMPLE 9

Simplify: $\left(\dfrac{x^2y}{xy^{-4}}\right)^3$.

Solution: $\dfrac{x^6y^3}{x^3y^{-12}} = \dfrac{x^6 y^3 y^{12}}{x^3} = x^{6-3}y^3y^{12} = x^3y^{15}$

Now try two problems yourself.

## Problem 1

Simplify: $\dfrac{x^9 y^{-6}}{x^{-3} y^2}$.

Solution:

Answer: $\dfrac{x^{12}}{y^8}$

## Problem 2

Simplify: $\left(\dfrac{2x^{-1}y^2}{x^{-3}}\right)^2$.

Solution:

Answer: $4x^4 y^4$

You should now be able to simplify expressions in which it is necessary to divide like variables with integral exponents. You also should be able to apply the definitions of integral exponents and to apply all five laws of exponents when simplifying expressions.

The classification scheme we developed for these laws, which are written to the right, should help you remember them.

Laws of Exponents

I. **Multiplication** $b^n \cdot b^m = b^{n+m}$      add exponents

II. **Power of a power** $(b^n)^m = b^{nm}$

III. **Power of a product** $(bc)^n = b^n c^n$         multiply exponents

IV. **Power of a fraction** $\left(\dfrac{a}{b}\right)^n = \dfrac{a^n}{b^n}$

V. **Division** $\dfrac{b^m}{b^n} = b^{m-n}$

or, alternatively,     subtract exponents

$= \dfrac{1}{b^{n-m}}$

Finally, I recommend establishing, and adhering to, your own pattern for simplification of expressions with exponents, such as mine. Before beginning the next unit you should simplify the expressions in the exercises. The final answers should be written without negative or zero exponents. The more of these problems you do, the easier they should become. When you have completed the exercises, check your answers against those given at the back of the book.

## EXERCISES

Simplify:

1. $\dfrac{5a^7 b^2}{ab^{10}}$

2. $w^5 \cdot w^0 \cdot w^{-7}$

3. $(3a^4 b^{-2})(a^5 b^{-3})$

4. $(4x^{-3} y^7)(-2x^2 y^2)$

5. $\dfrac{x^{-4}}{x^4}$

6. $\dfrac{15x^5 y^3}{3x^2 y^7}$

7. $\dfrac{x^5 \cdot x^{-4}}{x^{-3}}$

8. $x(3x^2 y^{-3})^2$

9. $(2w^{-2})^2 (5w^{-2})$

10. $x(5xy^{-2})^{-2}$

11. $\dfrac{m^{-9}s^{-8}}{m^{-4}s^3}$

12. $\dfrac{(3x^2y)^{-1}}{2xy^{-5}}$

13. $\dfrac{(3xy^{-2})^{-3}}{x}$

14. $\left[\dfrac{(3x^2y)^3}{3x^7y^9}\right]^2$

Hint: Since there are two sets of grouping symbols, work, as always, from the inside out to remove them.

15. $\left[\dfrac{(xy)^2}{x^{-1}}\right]^3$

16. $\left[\dfrac{(ab)^{-1}}{(a^{-2}b^3)^3}\right]^{-1}$   If you can do this one, you've got it made!

# UNIT 10

## Review of Fractions: Addition and Subtraction

In the preceding units we discussed positive integral exponents as well as zero and negative exponents. Before learning about **fractional** exponents, we need to review some facts about fractions themselves.

In this unit we will review how fractions are added and subtracted. When you have completed the unit, you will be able to solve problems involving the addition and subtraction of fractions, without changing the terms to decimals.

Before we begin, we must be certain of several definitions:

---

Definition:  Let $a$ and $b$ be integers with $b \neq 0$. Then $\dfrac{a}{b}$ is called a **rational number** (generally referred to as a **fraction**). In the fraction $\dfrac{a}{b}$, $a$ is the numerator and $b$ is the denominator.

---

Definition:  A **factor** is a number or letter that is being **multiplied**.

---

## EXAMPLE 1

Consider:  $3ax$. Since $3ax$ means $3 \cdot a \cdot x$, there are three factors: 3, $a$, and $x$.

---

> Definition: A fraction $\dfrac{a}{b}$ is said to be in **lowest terms** when all possible common factors **in the numerator and denominator** have been divided out.

## EXAMPLE 2

Consider: $\dfrac{6}{15} = \dfrac{2 \cdot \cancel{3}}{5 \cdot \cancel{3}} = \dfrac{2}{5};$ $\dfrac{2}{5}$ is in lowest terms.

---

Recall my earlier comments on the **sign** of a fraction.

$$\frac{-a}{b} = \frac{a}{-b} = -\frac{a}{b}$$

These three are equivalent fractions! The negative sign may be placed either in the numerator, in the denominator, or in front of the entire fraction. My preference is to place the negative sign in the numerator, which is what is done throughout this book.

Let me add a note of caution before proceeding. Please do *not* change the fractions to decimals, or you will defeat the intention of this and the following unit, which is to enable you to work effectively with algebraic as well as numerical fractions.

# ADDITION

> Definition: **Addition of Rationals with a Common Denominator**
>
> $$\frac{a}{b} + \frac{c}{b} = \frac{a+c}{b}$$

If the fractions have the same denominator, **add the numerators**. The denominator will remain the same.

## EXAMPLE 3

Add: $\dfrac{1}{7} + \dfrac{2}{7}$.

---

Solution: $\dfrac{1}{7} + \dfrac{2}{7} = \dfrac{3}{7}$

---

## EXAMPLE 4

Add:  $\dfrac{2}{5}+\dfrac{4}{5}$ .

Solution:  $\dfrac{2}{5}+\dfrac{4}{5}=\dfrac{2+4}{5}=\dfrac{6}{5}$

Note:   Leave the answer as an improper fraction; it is neither necessary nor advisable to change it to  $1\dfrac{1}{5}$ .

Unfortunately, few problems ever occur in which the fractions have the same denominator. However, the following definition allows us to add two fractions in one step without ever bothering to find the lowest common denominator. I rather like this particular definition, and I use it all the time.

Definition:   **Addition of Rationals with Unlike Denominators**

$$\frac{a}{b}+\frac{c}{d}=\frac{\textcircled{a}\,\textcircled{c}}{\textcircled{b}\,\textcircled{d}}=\frac{ad+bc}{bd}$$

Because of this definition, it is *not* necessary to find the least common denominator.

## EXAMPLE 5

Add:  $\dfrac{7}{8}+\dfrac{2}{3}$ .

Solution:  $\dfrac{7}{8}\times\dfrac{2}{3}=\dfrac{7\cdot3+8\cdot2}{8\cdot3}=\dfrac{21+16}{24}=\dfrac{37}{24}$

## EXAMPLE 6

Add:  $\dfrac{2}{3}+\dfrac{1}{5}$ .

Solution:  $\dfrac{2}{3}\times\dfrac{1}{5}=\dfrac{10+3}{15}=\dfrac{13}{15}$

Now try two problems yourself.

## Problem 1

Add: $\dfrac{4}{7}+\dfrac{5}{6}$.

Solution:

Answer: $\dfrac{59}{42}$

## Problem 2

Add: $\dfrac{-2}{3}+\dfrac{2}{5}$.

Solution:

Answer: $-\dfrac{4}{15}$

The advantage of the definition given above is that for simple fractions the addition can be performed in one step. However, it can be used only when adding two fractions; if there are more, the fractions must be added two at a time with this definition.

# SUBTRACTION

> Definition: **Subtraction of Rationals with Unlike Denominators**
>
> $$\frac{a}{b}-\frac{c}{d}=\frac{ad-bc}{bd}$$

Note that the only difference between this definition and the definition for the addition of rationals is the minus sign.

But be careful—the first term of the numerator must be the product found by multiplying diagonally to the right and **down** followed by the **minus sign** since it is a subtraction problem.

$$\frac{a}{b}\diagdown\frac{c}{d}=\frac{ad-}{\rule{2cm}{0.4pt}}$$

## EXAMPLE 7

Subtract: $\dfrac{3}{5} - \dfrac{2}{3}.$

Solution: $\dfrac{3}{5} \diagdown \dfrac{2}{3} = \dfrac{9-10}{15} = \dfrac{-1}{15}$

## EXAMPLE 8

Subtract: $\dfrac{-5}{11} - \dfrac{3}{4}.$

Solution: $\dfrac{-5}{11} \diagdown \dfrac{3}{4} = \dfrac{-20-33}{44} = \dfrac{-53}{44}$

## EXAMPLE 9

Subtract: $\dfrac{1}{3} - \dfrac{1}{2}.$

Solution: $\dfrac{1}{3} \diagdown \dfrac{1}{2} = \dfrac{2-3}{6} = \dfrac{-1}{6}$

## EXAMPLE 10

Solve: $\dfrac{1}{6} - \dfrac{-2}{3}.$

Solution: $\dfrac{1}{6} \diagdown \dfrac{-2}{3} = \dfrac{3-(-12)}{18} = \dfrac{3+12}{18} = \dfrac{15}{18} = \dfrac{3 \cdot 5}{3 \cdot 6} = \dfrac{5}{6}$

Note: After using the definition, our solution was not in the lowest terms and had to be reduced in the final step.

Try a couple of subtraction problems on your own. If you have forgotten the rules for adding and subtracting signed numbers, quickly review Unit 2 before proceeding.

## Problem 3

Solve: $\dfrac{5}{3} - \dfrac{1}{7}$.

Solution:

Answer: $\dfrac{32}{21}$

## Problem 4

Solve: $\dfrac{-3}{4} - \dfrac{2}{5}$.

Solution:

Answer: $\dfrac{-23}{20}$

With algebraic fractions the entire process becomes even easier.

## EXAMPLE 11

Add: $\dfrac{1}{x} + \dfrac{1}{y}$.

Solution: $\dfrac{1}{x} + \dfrac{1}{y} = \dfrac{y+x}{xy}$

## EXAMPLE 12

Add: $\dfrac{2x}{y} - \dfrac{z}{3}$.

---

Solution: $\dfrac{2x}{y} \searrow \dfrac{z}{3} = \dfrac{6x - yz}{3y}$

---

The next example illustrates how to add or subtract when only one of the terms is a fraction.

## EXAMPLE 13

Subtract: $x - \dfrac{1}{2}$.

---

Solution: $x - \dfrac{1}{2} = \dfrac{x}{1} - \dfrac{1}{2} = \dfrac{x}{1} \searrow \dfrac{1}{2} = \dfrac{2x - 1}{2}$

First the $x$ was rewritten as $\dfrac{x}{1}$, and then the two fractions were subtracted.

---

We will conclude this unit with one final example.

## EXAMPLE 14

Subtract: $\dfrac{x+2}{5} - \dfrac{x-1}{7}$.

---

Solution: $\dfrac{x+2}{5} \searrow \dfrac{x-1}{7} = \dfrac{7(x+2) - 5(x-1)}{35}$

$$= \dfrac{7x + 14 - 5x + 5}{35}$$

$$= \dfrac{2x + 19}{35}$$

---

You should now be able to add and subtract rationals (fractions). Before beginning the next unit, you should solve the following problems involving rationals. Reduce the answers to the lowest terms, but *do not change them to decimals.*

## EXERCISES

Solve:

1. $\dfrac{2}{11} + \dfrac{1}{11}$

2. $\dfrac{7}{10} - \dfrac{9}{10}$

3. $\dfrac{7}{9} + \dfrac{1}{5}$

4. $\dfrac{1}{x} + 5$     Hint: Rewrite 5 as $\dfrac{5}{1}$.

5. $3 - \dfrac{5}{w}$

6. $7 + \dfrac{1}{x}$

7. $\dfrac{2}{9} - \dfrac{-1}{10}$

8. $\dfrac{11}{t} + \dfrac{7}{r}$

9. $\dfrac{1}{x} + \dfrac{1}{x}$

10. $\dfrac{10}{x+1} + \dfrac{3}{x+1}$

11. $\dfrac{5}{a} - \dfrac{4}{a}$

12. $\dfrac{-s}{9} + \dfrac{k}{10}$

13. $\dfrac{x}{2} - \dfrac{x}{5}$

14. $\dfrac{x+1}{2} - \dfrac{3}{5}$

15. $\dfrac{x-1}{3} + \dfrac{x+1}{2}$

16. $\dfrac{x+2}{2} - \dfrac{x+3}{3}$

# UNIT 11

## Review of Fractions: Multiplication and Division; Complex Fractions

Here, as in Unit 10, we will review some facts about **fractions** before we study **fractional exponents**. When you have completed this unit, you will be able to solve problems involving the multiplication and division of fractions, without changing the terms to decimals.

## MULTIPLICATION

Definition:  **Multiplication of Rationals**

$$\frac{a}{b} \cdot \frac{c}{d} = \frac{ac}{bd}$$

Multiply numerators together, and multiply denominators together.

Note:  1.  You do not need a common denominator.
2.  You should factor and cancel as soon as possible.

## EXAMPLE 1

Multiply: $\dfrac{2}{3} \cdot \dfrac{4}{5}$.

---

Solution: $\dfrac{2}{3} \cdot \dfrac{4}{5} = \dfrac{2 \cdot 4}{3 \cdot 5} = \dfrac{8}{15}$

---

## EXAMPLE 2

Multiply: $\dfrac{4}{7} \cdot \dfrac{35}{12}$.

---

Solution:
$$\overset{1}{\underset{1}{\cancel{4}}} \cdot \frac{\overset{5}{\cancel{35}}}{\underset{3}{\cancel{12}}} \cdot \overset{}{} = \frac{5}{3}$$

$$\frac{\overset{1}{\cancel{4}}}{\underset{1}{\cancel{7}}} \cdot \frac{\overset{5}{\cancel{35}}}{\underset{3}{\cancel{12}}} = \frac{5}{3}$$

---

## EXAMPLE 3

Multiply: $\dfrac{12}{15} \cdot \dfrac{5}{18}$.

---

Solution:
$$\frac{\overset{2}{\cancel{12}}}{\underset{3}{\cancel{15}}} \cdot \frac{\overset{1}{\cancel{5}}}{\underset{3}{\cancel{18}}} = \frac{2 \cdot 1}{3 \cdot 3} = \frac{2}{9}$$

---

# DIVISION

| Definition: **Division of Rationals** |
|---|
| $\dfrac{a}{b} \div \dfrac{e}{f} = \dfrac{a}{b} \cdot \dfrac{f}{e} = \dfrac{af}{be}$ |

When dividing by a fraction, we invert the second fraction (the divisor) and multiply. Or, another way of saying it, we multiply by the reciprocal of the divisor.

## EXAMPLE 4

Divide:   $\dfrac{2}{3} \div \dfrac{-3}{8}$.

Solution:   $\dfrac{2}{3} \div \dfrac{-3}{8} = \dfrac{2}{3} \cdot \dfrac{8}{-3} = \dfrac{16}{-9} = \dfrac{-16}{9}$

As stated in the previous unit, with algebraic fractions the entire process becomes even easier.

## EXAMPLE 5

Multiply:   $\dfrac{a}{b} \cdot \dfrac{5}{a}$.

Solution:   $\dfrac{a}{b} \cdot \dfrac{5}{a} = \dfrac{\cancel{a} \cdot 5}{b \cdot \cancel{a}} = \dfrac{5}{b}$

## EXAMPLE 6

Divide:   $\dfrac{x+1}{3} \div 2$.

Solution:   $\dfrac{x+1}{3} \div 2 = \dfrac{x+1}{3} \cdot \dfrac{1}{2} = \dfrac{(x+1) \cdot 1}{3 \cdot 2} = \dfrac{x+1}{6}$

Here are four problems for you.

## Problem 1

Simplify:   $\dfrac{2}{-5} \cdot \dfrac{30}{8}$.

Solution:

Answer:   $-\dfrac{3}{2}$

## Problem 2

Simplify:  $\dfrac{2}{-5} \div \dfrac{30}{8}$.

Solution:

Answer:  $-\dfrac{8}{75}$

## Problem 3

Simplify:  $\dfrac{-12}{20} \div \dfrac{10}{-16}$.

Solution:

Answer:  $\dfrac{24}{25}$

## Problem 4

Divide:  $\dfrac{w+x}{2} \div \dfrac{1}{2}$.

Solution:

Answer:  $w + x$

# COMPLEX FRACTIONS

We must now consider **complex fractions**—fractions in which there are one or more fractions in the numerator or denominator, or in both. Remember that in a complex fraction, as in a simple fraction, the horizontal bar means simply that we should divide. For example:

$$\frac{12}{2} \text{ means 12 divided by 2}$$

Therefore, when I am faced with a complex fraction, the first thing I do is to rewrite it as a simple division problem. Remember that if the numerator or the denominator, or both, have more than one term, you will need to enclose them in parentheses when rewriting the complex fraction as a division problem.

## EXAMPLE 7

Simplify this complex fraction: $\dfrac{\frac{2}{5}}{w}$.

Solution: $\dfrac{\frac{2}{5}}{w} = \frac{2}{5} \div w$     Rewrite, using a division sign.

$$= \frac{2}{5} \cdot \frac{1}{w}$$

$$= \frac{2}{5w}$$

## EXAMPLE 8

Simplify: $\dfrac{\frac{1}{2} + \frac{2}{3}}{\frac{5}{4}}$.

Solution: $\dfrac{\frac{1}{2}+\frac{2}{3}}{\frac{5}{4}} = \left(\dfrac{1}{2} + \dfrac{2}{3}\right) \div \dfrac{5}{4}$  Rewrite, using a division sign.

$$= \left(\dfrac{3+4}{6}\right) \div \dfrac{5}{4} \qquad \text{Add fractions in parentheses.}$$

$$= \dfrac{7}{6} \cdot \dfrac{4}{5}$$

$$= \dfrac{7}{\overset{}{\underset{3}{\cancel{6}}}} \cdot \dfrac{\overset{2}{\cancel{4}}}{5}$$

$$= \dfrac{14}{15}$$

---

## EXAMPLE 9

Simplify: $\dfrac{3+\frac{1}{x}}{2}$.

---

Solution: $\dfrac{3+\frac{1}{x}}{2} = \left(3 + \dfrac{1}{x}\right) \div 2$  Rewrite, using a division sign.

$$= \left(\dfrac{3}{1} + \dfrac{1}{x}\right) \div \dfrac{2}{1}$$

$$= \dfrac{3x+1}{x} \cdot \dfrac{1}{2} \qquad \text{Add fractions in parentheses.}$$

$$= \dfrac{3x+1}{2x}$$

---

Can you do this one?

---

## Problem 5

Simplify: $\dfrac{\frac{3}{x}}{1+\frac{7}{x}}$

---

Solution:

Answer: $\dfrac{3}{x+7}$

---

Here is one more example to end the unit.

## EXAMPLE 10

Simplify: $\dfrac{5-\dfrac{3}{x}}{x}$.

Solution:
$$\dfrac{5-\dfrac{3}{x}}{x} = \left(\dfrac{5}{1} - \dfrac{3}{x}\right) \div x$$
$$= \left(\dfrac{5x-3}{x}\right) \cdot \dfrac{1}{x}$$
$$= \dfrac{(5x-3) \cdot 1}{x \cdot x}$$
$$= \dfrac{5x-3}{x^2}$$

You should now be able to multiply and divide fractions. You should also be able to simplify complex fractions in which there are fractions in the numerator and/or denominator.

Before beginning the next unit, do the following exercises, reducing the answers to the lowest terms without *converting to decimals*.

## EXERCISES

Simplify:

1. $\dfrac{4}{7} \cdot \dfrac{35}{12}$

2. $\dfrac{5}{18} \div \dfrac{3}{14}$

3. $\dfrac{3}{4} \cdot \dfrac{-8}{9}$

4. $\dfrac{9}{14} \div \dfrac{5}{21}$

5. $\dfrac{-8}{9} \div \dfrac{12}{-7}$

6. $\left(\dfrac{2}{x} \cdot \dfrac{x}{5}\right) \div w$

7. $\dfrac{\dfrac{3}{10}}{\dfrac{1}{10}}$

8. $\dfrac{3-\dfrac{2}{5}}{3+\dfrac{2}{5}}$

9. $\dfrac{1-\dfrac{1}{3}}{\dfrac{5}{6}}$

10. $\dfrac{\dfrac{a}{2}-\dfrac{3}{5}}{2}$

11. $7 \div \left(\dfrac{x-1}{4}\right)$

12. $\dfrac{\dfrac{2}{x}-5}{x}$

13. $\dfrac{9-\dfrac{x}{4}}{\dfrac{1}{2}}$

14. $\dfrac{x+\dfrac{x+2}{2}}{\dfrac{x}{2}}$

15. $\dfrac{\dfrac{a}{b}+2}{\dfrac{a}{b}+1}$

# UNIT 12

# Square Roots and Radicals

In this unit you will learn what is meant by the root of a number or algebraic expression and the vocabulary used with radicals. When you have completed the unit, you will be able to multiply, divide, add, and subtract radical expressions, both numerical and algebraic.

## ROOTS

The **square root** of a number or expression is one of its *two equal factors*. For example, +5 is a square root of 25 since $(+5)(+5) = 25$. Also, −5 is a square root of 25 since $(−5)(−5) = 25$.

A positive number has two square roots, which are opposites of each other. To indicate both square roots, the symbol $\pm$ may be used. Thus the square roots of 49 are $\pm 7$.

The **cube root** of a number or expression is one of its *three equal factors*. For example, 2 is the cube root of 8 since $(2)(2)(2) = 8$, and −3 is the cube root of −27 since $(−3)(−3)(−3) = −27$, and $2x$ is the cube root of $8x^3$ since $(2x)(2x)(2x) = 8x^3$.

The **fourth root** of a number is one of its *four equal factors*, and so on for fifth roots, etc.

---

EXAMPLE 1   The square roots of 36 are $\pm 6$ since $(6)(6) = 36$ and $(−6)(−6) = 36$.

EXAMPLE 2   The cube root of −8 is −2 since $(−2)(−2)(−2) = −8$.

EXAMPLE 3   The fifth root of 1 is 1 since $(1)(1)(1)(1)(1) = 1$.

EXAMPLE 4   The square root of −16 does not exist in the real number system because there is no real number such that when it is squared will be a negative number.

---

From the four examples we should note the following:

1. A positive number has two square roots, which are opposites of each other.

2. A positive number has only one cube root.

3. The cube root of a negative number exists and is negative.

4. The square root of a negative number does not exist in the set of real numbers.

Try the next four problems on your own.

---

**Problem 1**   Find the square roots of 64.

**Problem 2**   Find the square roots of 100.

**Problem 3**   Find the cube root of –1.

**Problem 4**   Find the square root of 0.

Answers:   1. ±8   2. ±10   3. –1   4. 0

---

# RADICALS

A **radical** is an indicated root of a number or expression such as $\sqrt{5}$, $\sqrt[3]{8x}$, and $\sqrt[4]{7x^3}$. The symbol, $\sqrt{\phantom{x}}$, is called a **radical sign**.

The **radicand** is the number or expression under the radical sign. In the above examples the radicands are 5, 8x, and $7x^3$.

The **index** is the small number written above and to the left of the radical sign. The index indicates which root is to be taken. In square roots, the index 2 is not written. In the above examples the indices are 2, 3, and 4.

The **principal square root** of a number is its positive square root. The radical sign is used to indicate the principal square root. Thus the square roots of 49 are ±7, but $\sqrt{49}$ = principal square root of 49 = +7. To indicate the negative square root of a number, a negative sign is placed before the symbol. $-\sqrt{16} = -4$.

By definition, $\sqrt{0} = 0$.

---

## EXAMPLE 5

Find $\sqrt{4}$.

---

Solution:   The problem is asking for the principal square root of 4, which is the positive root only.

Answer:   $\sqrt{4} = 2$

---

## EXAMPLE 6

Find $\sqrt[3]{-125}$.

---

Solution:   The problem is asking for the cube root of $-125$.

Answer:   $\sqrt[3]{-125} = -5$ since $(-5)(-5)(-5) = -125$

---

Numbers such as $\sqrt{5}$, $\sqrt{23}$, and $\sqrt[3]{17}$, where the radicand is not a square or a cube, are irrational numbers. If a decimal approximation is needed, it can be found by using a calculator; otherwise the radical is left as written. The intent in this and later units is to learn to work with radical expressions, not their decimal approximations.

## SQUARE ROOTS OF SQUARES

For the remainder of this book, we will assume that all letters—$a$, $b$, $c, \ldots,$ $x$, $y$, $z$— that appear under a radical sign represent positive real numbers unless otherwise specified. Thus $\sqrt{x^2} = x$ since $x$ is assumed to be positive and is the principal square root. In later mathematics courses, you will need to consider the possibility that $x$ is negative, and the principal square root becomes $\sqrt{x^2} = |x|$.

In the following examples the algebraic expression under the radical sign, the radicand, is a square. To find the square root of a monomial, which is a square, we need only keep the base the same and take one half of the exponent.

---

**EXAMPLE 7**     $\sqrt{y^2} = y$ since $y \cdot y = y^2$

**EXAMPLE 8**     $\sqrt{9w^2} = 3w$ since $(3w)(3w) = (3w)^2 = 3^2 w^2$

**EXAMPLE 9**     $\sqrt{a^2 c^2} = ac$ since $(ac)(ac) = (ac)^2 = a^2 c^2$

**EXAMPLE 10**    $\sqrt{x^6} = x^3$ since $x^3 \cdot x^3 = x^6$

**EXAMPLE 11**    $\sqrt{z^{10}} = z^5$ since $z^5 \cdot z^5 = z^{10}$

---

## MULTIPLICATION AND DIVISION

Now that you are able to simplify square radicands, we will consider multiplying and dividing square roots.

$$\boxed{\text{Rule:}\quad \sqrt{a} \cdot \sqrt{b} = \sqrt{ab}}$$

In words, the rule states that the product of square roots is the square root of the radicands. If the resulting new radicand is a square, it should be simplified as illustrated in the following examples.

**EXAMPLE 12**   $\sqrt{2}\cdot\sqrt{5}=\sqrt{2\cdot5}=\sqrt{10}$

**EXAMPLE 13**   $\sqrt{5}\cdot\sqrt{x}=\sqrt{5\cdot x}=\sqrt{5x}$

**EXAMPLE 14**   $\sqrt{a^3}\cdot\sqrt{a}=\sqrt{a^3\cdot a}=\sqrt{a^4}=a^2$

**EXAMPLE 15**   $\sqrt{2x}\cdot\sqrt{8x}=\sqrt{2\cdot8\cdot x\cdot x}=\sqrt{16x^2}=4x$

**EXAMPLE 16**   $\sqrt{5}\left(\sqrt{3}+\sqrt{w}\right)=\sqrt{5\cdot3}+\sqrt{5\cdot w}=\sqrt{15}+\sqrt{5w}$

A similar rule applies for dividing square root radicals.

$$\text{Rule:}\quad \frac{\sqrt{a}}{\sqrt{b}}=\sqrt{\frac{a}{b}}$$

In words, the rule states that the quotient of two square roots is equal to the square root of the radicands. One example should be sufficient.

**EXAMPLE 16**   $\dfrac{\sqrt{27x^3}}{\sqrt{3x}}=\sqrt{\dfrac{27x^3}{3x}}=\sqrt{9x^2}=3x$

It's time for you to try a few problems on your own. Perform the indicated operations and simplify the square roots of any squares.

**Problem 5**   $\sqrt{49g^2}$

**Problem 6**   $\sqrt{20w}\cdot\sqrt{5w^3}$

**Problem 7**   $\sqrt{x}\cdot\sqrt{x}$

Answers:   5. $7g$   6. $10w^2$   7. $x$

# SIMPLIFYING SQUARE ROOTS OF POWERS

This section deals with how to simplify an expression like $\sqrt{x^3}$, where the expression under the radical is not itself a square but does have square factors. We can use the rule for multiplying square roots in reverse. First rewrite the expression as a product of the greatest square factor times the remaining factors. To simplify, take the square root of the square.

EXAMPLE 17 $\quad \sqrt{x^3} = \sqrt{x^2 \cdot x} = \sqrt{x^2} \cdot \sqrt{x} = x\sqrt{x}$

EXAMPLE 18 $\quad \sqrt{12x} = \sqrt{4 \cdot 3x} = \sqrt{4} \cdot \sqrt{3x} = 2\sqrt{3x}$

EXAMPLE 19 $\quad \sqrt{50w^5} = \sqrt{25 \cdot 2 \cdot w^4 \cdot w} = \sqrt{25w^4} \cdot \sqrt{2w} = 5w^2\sqrt{2w}$

# ADDITION AND SUBTRACTION

To add or subtract radicals, the radicands must be the same. Then combine like terms using the distributive property, $a(b + c) = ab + ac$, as shown in the next two examples.

EXAMPLE 20 $\quad 2\sqrt{7} + 3\sqrt{7} = (2+3)\sqrt{7} = 5\sqrt{7}$

EXAMPLE 21 $\quad \sqrt{2x} + 9\sqrt{2x} = (1+9)\sqrt{2x} = 10\sqrt{2x}$

The final example requires that the radicals be simplified before attempting to do the addition.

## EXAMPLE 22

Simplify: $\sqrt{32w} + \sqrt{18w}$.

Solution: $\quad \sqrt{32w} + \sqrt{18w} = \sqrt{16 \cdot 2w} + \sqrt{9 \cdot 2w}$

$$= 4\sqrt{2w} + 3\sqrt{2w}$$
$$= (4+3)\sqrt{2w}$$
$$= 7\sqrt{2w}$$

You should now be able to simplify, add, subtract, multiply, and divide square root radicals, both numerical and algebraic expressions.

Before continuing on to the next unit, you should simplify the following radicals. Do not use decimal approximations but instead work with simplifying the expressions written as radicals.

## EXERCISES

Simplify:

1. $\sqrt{5} \cdot \sqrt{20}$

2. $\sqrt{75}$

3. $\sqrt{2a} \cdot \sqrt{3b}$

4. $\sqrt{0}$

5. $\sqrt{3} \cdot \sqrt{6}$

6. $\sqrt{64t^2}$

7. $\sqrt{w^4}$

8. $\sqrt{45}$

9. $\sqrt{\dfrac{25}{x^2}}$

10. $\dfrac{\sqrt{20}}{\sqrt{5}}$

11. $11\sqrt{2} + 3\sqrt{2}$

12. $\sqrt{3x} \cdot \sqrt{3x}$

13. $7\sqrt{40} - 2\sqrt{10}$

14. $\sqrt{12y^8}$

15. $\sqrt{50x^4}$

16. $\sqrt{3}(\sqrt{2} + 1)$

17. $\sqrt{5}(\sqrt{5} + \sqrt{3})$

18. $\sqrt{2a^2c} \cdot \sqrt{2ac}$

# UNIT 13

## Fractional Exponents

In Units 10 and 11 we reviewed how to add, subtract, multiply, and divide fractions. Then in Unit 12 you learned what is meant by the root of a number or algebraic expression and the vocabulary used with radicals. Now we are ready to proceed to handling the most difficult exponents, fractional exponents.

First recall how we define integral exponents:

| | |
|---|---|
| Positive Integer: | $b^n = \underbrace{b \cdot b \cdot b \cdot \ldots \cdot b}_{n \text{ factors}}$ |
| Zero: | $b^0 = 1$ |
| Negative: | $b^{-n} = \dfrac{1}{b^n}$ and $\dfrac{1}{b^{-n}} = b^n$ |

Now we will define a fractional exponent.

| |
|---|
| Definition: **Fractional Exponent** |
| $b^{n/d} = \left(\sqrt[d]{b}\right)^n = \sqrt[d]{b^n}$ |
| if $d$ is even, $b$ must be nonnegative. |

Using the vocabulary introduced in Unit 12, the letter $d$ is called the **index** of the radical and $b$ is called the **radicand**.

Note: $b^{n/d}$ — denotes power / denotes the root

Note that two forms of the definition are given.

$$b^{n/d} = \left(\sqrt[d]{b}\right)^n = \sqrt[d]{b^n}$$

numerical  algebraic

The first form is useful in numerical calculations, provided that the $d$th root of $b$ is a known integer. It is then convenient to take the root before raising to the $n$th power in order to work with smaller numbers.

The second form is the more common way of rewriting algebraic expressions with fractional exponents.

As the two forms suggest, we can do either the root or the power first. Here we will consider some numerical problems first.

### Fractional Exponent with Numerical Problems

power  root

$$b^{n/d} = \left(\sqrt[d]{b}\right)^n$$

numerical

**EXAMPLE 1**  $8^{1/3} = \left(\sqrt[3]{8}\right)^1 = (2)^1 = 2$

**EXAMPLE 2**  $4^{1/2} = \left(\sqrt{4}\right)^1 = (2)^1 = 2$

Remember that the 2 is not usually written for square roots.

**EXAMPLE 3**  $4^{3/2} = \left(\sqrt{4}\right)^3 = (2)^3 = 8$

**EXAMPLE 4**  $27^{2/3} = \left(\sqrt[3]{27}\right)^2 = (3)^2 = 9$

**EXAMPLE 5**  $8^{4/3} = \left(\sqrt[3]{8}\right)^4 = (2)^4 = 16$

**EXAMPLE 6**  $9^{-1/2} = \left(\sqrt{9}\right)^{-1} = (3)^{-1} = \dfrac{1}{3}$

or, alternatively,

$$= \frac{1}{9^{1/2}} = \frac{1}{\sqrt{9}} = \frac{1}{3}$$

It is important that you learn the meaning of, and be able to work with, fractional exponents. Punching numbers into a calculator defeats this purpose. So for now try to simplify the following numerical problems with fractional exponents without using a calculator. You should be able to find the various roots easily.

Problem 1    $32^{1/5} =$

Problem 2    $(-1)^{1/3} =$

Problem 3    $4^{5/2} =$

Problem 4    $64^{2/3} =$

Problem 5    $81^{3/4} =$

Problem 6    $(-8)^{2/3} =$

Answers:   1. 2    2. −1    3. 32    4. 16    5. 27    6. 4

And now for some problems where the root is not a rational number. When the root is not a rational number, as in the next two examples, the expression is rewritten as a radical. Although the two expressions are equivalent, the radical form is considered to be the simplified form. For example, $2^{1/2}$ and $\sqrt{2}$ are equivalent, but $\sqrt{2}$ is considered to be the simplified form.

**EXAMPLE 7**    Rewrite as a radical:    $5^{2/3} = \sqrt[3]{5^2} = \sqrt[3]{25}$

**EXAMPLE 8**    Rewrite as a radical:    $7^{1/5} = \sqrt[5]{7}$

## Fractional Exponent with Algebraic Expressions

$$\overset{\text{power}\quad\text{root}}{b^{n/d} = \sqrt[d]{b^n}}$$
$$\underset{\text{algebraic}}{}$$

**EXAMPLE 9**    Rewrite as a radical:    $x^{2/3} = \sqrt[3]{x^2}$

**EXAMPLE 10**    Rewrite as a radical:    $y^{2/5} = \sqrt[5]{y^2}$

The only time any simplifying can be done is if $n/d$ is an improper fraction. Given an algebraic expression with an improper fractional exponent, rewrite the expression as a radical and simplify as explained in Unit 12. The following examples illustrate the procedure.

EXAMPLE 11     $x^{3/2} = \sqrt{x^3} = \sqrt{x^2 \cdot x} = x\sqrt{x}$

The final answer is preferred to $\sqrt{x^3}$.

EXAMPLE 12     $x^{7/5} = \sqrt[5]{x^7} = \sqrt[5]{x^5 \cdot x^2} = x\sqrt[5]{x^2}$

EXAMPLE 13     $x^{4/3} = \sqrt[3]{x^4} = \sqrt[3]{x^3 \cdot x} = x\sqrt[3]{x}$

EXAMPLE 14     $x^{5/2} = \sqrt{x^5} = \sqrt{x^4 \cdot x} = x^2\sqrt{x}$

---

## ALGEBRA AND THE CALCULATOR (Optional)

In some instances you may need a decimal approximation for a radical. There is a calculator key for square roots, which makes these problems relatively easy. But what about examples such as $5^{2/3}$ and $-10^{-1/5}$? The $\boxed{\wedge}$ key is used to denote exponentiation. If the exponent is a fraction, parentheses must be used to enclose the entire fraction. In the next two examples the keystrokes are shown for finding the decimal approximations.

EXAMPLE 15     Using a calculator, find the decimal approximation for $5^{2/3}$.

Solution:   To find the answer, press $\boxed{5}\ \boxed{\wedge}\ \boxed{(}\ \boxed{2}\ \boxed{\div}\ \boxed{3}\ \boxed{)}\ \boxed{\text{ENTER}}$.

Answer:   2.924, correct to three decimal places

EXAMPLE 16     Using a calculator, find the decimal approximation for $-10^{-1/5}$.

Solution:   To find the answer, press $\boxed{(-)}\ \boxed{1}\ \boxed{0}\ \boxed{\wedge}\ \boxed{(}\ \boxed{(-)}\ \boxed{1}\ \boxed{\div}\ \boxed{5}\ \boxed{)}\ \boxed{\text{ENTER}}$.

Answer:   −0.631, correct to three decimal places

---

You should now be able to simplify expressions involving fractional exponents, whether they occur in numerical or in algebraic problems. Note that we are concerned mainly with the ability to rewrite fractional exponents using radicals and are *not* concerned about using a calculator to find decimal approximations.

Before beginning the next unit, you should simplify the following expressions involving fractional exponents.

# EXERCISES

Simplify the following without using a calculator:

1.  $32^{1/5}$

2.  $(-1)^{2/3}$

3.  $(-4)^{1/2}$    Careful!

4.  $4^{3/2}$

5.  $4^{1/2}$

6.  $4^{-1/2}$

Rewrite the following expressions using radicals. Simplify whenever possible.

7.  $x^{-1/2}$

8.  $x^{1/3}$

9.  $a^{2/5}$

10.  $4^{-3/2}$

11.  $(x + 1)^{1/2}$

12.  $x^{8/3}$

13.  $(4x)^{1/2}$

14.  $x^{11/2}$

15.  $(5x)^{-1/2}$

16.  $(18x^3)^{1/2}$

17.  $(2x)^{2/3}$

18.  $(-64)^{2/3}$

19.  Rewrite using a fractional exponent: $\sqrt{7x}$.

20.  Rewrite using a fractional exponent: $\sqrt[3]{2x}$.

C21.  Find the decimal approximation correct to three decimal places for $23^{1/5}$.

C22.  Find the decimal approximation correct to three decimal places for $\sqrt[4]{17}$.

# UNIT 14

# Simplifying Expressions with Fractional Exponents

You have completed all the definitions and laws of exponents necessary to handle any given expression involving exponents. From here on, you need practice, confidence, and a bit of luck! We're going to work on practice in this unit.

The definitions and laws of exponents are restated below for reference:

---

Positive Integer: $b^n = \underbrace{b \cdot b \cdot b \cdot \ldots \cdot b}_{n \text{ factors}}$

Zero: $b^0 = 1$

Negative: $b^{-n} = \dfrac{1}{b^n}$ and $\dfrac{1}{b^{-n}} = b^n$

Fractional: $b^{n/d} = \sqrt[d]{b^n} = \left(\sqrt[d]{b}\right)^n$

---

Laws of Exponents

I.   **Multiplication** $\quad b^n \cdot b^m = b^{n+m}$ $\qquad\qquad$ add exponents

II.  **Power of a power** $\quad (b^n)^m = b^{nm}$

III. **Power of a product** $\quad (bc)^n = b^n c^n$ $\qquad\qquad$ multiply exponents

IV.  **Power of a fraction** $\quad \left(\dfrac{a}{b}\right)^n = \dfrac{a^n}{b^n}$

V.   **Division** $\qquad\qquad \dfrac{b^m}{b^n} = b^{m-n}$

$\qquad\qquad\qquad\qquad$ or, alternatively, $\qquad\qquad$ subtract exponents

$\qquad\qquad\qquad\qquad = \dfrac{1}{b^{n-m}}$

Here we get into some dandy problems. Luckily, fractional exponents obey all the laws of integer exponents.

To simplify expressions involving exponents means to write the final answer without using zero, negative, or fractional exponents. To accomplish this, as in previous units, I **always** will use the same procedure when simplifying. Of course, once again an additional step has been added, this time to include fractional exponents.

My suggested procedure for simplifying an algebraic expression with exponents is:

- If there are parentheses, remove them using Laws II, III, and IV (multiply exponents).
- If there are negative exponents, use the definition to rewrite them as positive exponents.
- If applicable, use the division law, Law V (subtract the exponents).
- If applicable, use the multiplication law, Law I (add the exponents).
- If fractional exponents remain, use the definition to rewrite as a radical.

It is the same basic pattern I have been using since Unit 7.

---

## EXAMPLE 1

Simplify: $\left(x^{-1/2}\right)^{-2/3}$.

---

Solution: $\left(x^{-1/2}\right)^{-2/3} = x^{1/3}$      Remove parentheses; multiply exponents.

$$\frac{-1}{2} \cdot \frac{-2}{3} = \frac{\overset{1}{\cancel{2}}}{\underset{1}{\cancel{2} \cdot 3}} = \frac{1}{3}$$

$\qquad\qquad = \sqrt[3]{x}$      Rewrite as a radical.

---

## EXAMPLE 2

Simplify: $x^{-1/3} \cdot x^{1/2}$.

---

Solution: $x^{-1/3} \cdot x^{1/2} = x^{1/6}$      Use the multiplication law; add exponents.

$$\frac{-1}{3} \,\diagdown\, \frac{1}{2} = \frac{-2+3}{6} = \frac{1}{6}$$

$\qquad\qquad = \sqrt[6]{x}$      Rewrite as a radical.

---

## EXAMPLE 3

Simplify:   $(9x^{-4})^{1/2}$.

Solution:   $(9x^{-4})^{1/2} = 9^{1/2} \cdot x^{-2}$   Remove parentheses; multiply exponents.

$$1 \cdot \frac{1}{2} = \frac{1}{2} \text{ and } (-4) \cdot \frac{1}{2} = \frac{\overset{2}{\cancel{-4}}}{1} \cdot \frac{1}{\underset{1}{\cancel{2}}} = -2$$

$= \dfrac{9^{1/2}}{x^2}$      Rewrite with positive exponents.

$= \dfrac{3}{x^2}$      since $9^{1/2} = \sqrt{9} = 3$.

## EXAMPLE 4

Simplify:   $(x^6 y^{-3})^{-2/3}$.

Solution:   $(x^6 y^{-3})^{-2/3} = x^{-4} y^2$   Remove parentheses; multiply exponents.

$$\frac{\overset{2}{\cancel{6}}}{1} \cdot \frac{-2}{\underset{1}{\cancel{3}}} = -4 \text{ and } \frac{\overset{1}{\cancel{-3}}}{1} \cdot \frac{-2}{\underset{1}{\cancel{3}}} = 2$$

$= \dfrac{y^2}{x^4}$      Rewrite with positive exponents.

You try one; then we'll have some more examples.

## Problem 1

Simplify:    $(x^{-8} y^4)^{-3/2}$.

Solution:

Answer:   $\dfrac{x^{12}}{y^6}$

## EXAMPLE 5

Simplify: $\left(\dfrac{a^{1/2}b^{2/3}}{c^{1/7}}\right)^6$.

---

Solution: $\left(\dfrac{a^{1/2}b^{2/3}}{c^{1/7}}\right)^6 = \dfrac{a^3 b^4}{c^{6/7}}$   Remove parentheses; multiply exponents.

$$\frac{1}{\cancel{2}}\cdot\frac{\overset{3}{\cancel{6}}}{1}=3,\quad \frac{2}{\cancel{3}}\cdot\frac{\overset{2}{\cancel{6}}}{1}=4 \text{ and } \frac{1}{7}\cdot\frac{6}{1}=\frac{6}{7}$$

$= \dfrac{a^3 b^4}{\sqrt[7]{c^6}}$   Rewrite fractional exponent as a radical.

At this point I'm not at all concerned about rationalizing that denominator. Recall from high school days that it was a "no-no" to leave the radical in the denominator.

---

Quite often expressions you are trying to simplify contain radicals. Begin by using the definition of a fractional exponent to rewrite the radical expression using exponents. This requires adding one final step to my suggested procedure for simplifying algebraic expressions involving exponents.

1. If there are radicals, use the definition to rewrite them as fractional exponents.

2. If there are parentheses, remove them using Laws II, III, and IV (multiply exponents).

3. If there are negative exponents, use the definition to rewrite them as positive exponents.

4. If applicable, use the division law of exponents, Law V (subtract exponents).

5. If applicable, use the multiplication law of exponents, Law I (add exponents).

6. If fractional exponents remain, use the definition to rewrite them as radicals.

You will be happy to learn that in the majority of the problems encountered in this book, it will not be necessary to use all six steps. Often one or two steps become unnecessary, but the order of the steps **always** remains the same in my examples.

## EXAMPLE 6

Simplify: $\sqrt{x}\cdot\sqrt[3]{x}$.

---

Solution: $\sqrt{x}\cdot\sqrt[3]{x} = x^{1/2}\cdot x^{1/3}$   Rewrite the radicals using fractional exponents.

$= x^{5/6}$   Use the multiplication law; add exponents.

$$\frac{1}{2}+\frac{1}{3}=\frac{3+2}{6}=\frac{5}{6}$$

$= \sqrt[6]{x^5}$   Rewrite the fractional exponent as a radical.

---

## EXAMPLE 7

Simplify: $\dfrac{\sqrt[3]{x^2}}{\sqrt[4]{x}}$.

Solution: $\dfrac{\sqrt[3]{x^2}}{\sqrt[4]{x}} = \dfrac{x^{2/3}}{x^{1/4}}$    Rewrite radicals using fractional exponents.

$= x^{5/12}$    Use the division law; subtract exponents.

$$\dfrac{2}{3} \searrow \dfrac{1}{4} = \dfrac{8-3}{12} = \dfrac{5}{12}$$

$= \sqrt[12]{x^5}$    Rewrite the fractional exponent as a radical.

## EXAMPLE 8

Simplify: $\left[\left(\dfrac{\sqrt{2}}{-3}\right)^{-4}\right]^{-1}$.

Solution: $\left[\left(\dfrac{(2)^{1/2}}{-3}\right)^{-4}\right]^{-1} = \left[\dfrac{(2)^{-2}}{(-3)^{-4}}\right]^{-1} = \dfrac{2^2}{(-3)^4} = \dfrac{4}{81}$

As always, work from the inside out to remove parentheses.

Here are four problems for you.

## Problem 2

Simplify: $\left(\sqrt[3]{7}\right)^6$.

Solution:

Answer:    49

## Problem 3

Simplify: $\left(x^{1/3}\right)^3$.

Solution:

Answer:  $x$

## Problem 4

Simplify: $\left(\dfrac{2^0}{8^{1/3}}\right)^{-1}$.

Solution:

Answer:  2

Did you get the correct answer to Problem 4? If not, maybe you forgot that when removing the parentheses, you multiply the exponents, which yields: $0 \cdot (-1) = 0$ and then $2^0 = 1$.

## Problem 5

Simplify: $\left(2\sqrt[6]{x}\right)^3$.

Solution:

Answer:  $8\sqrt{x}$

In case you had difficulty with Problems 2–5, let me restate our objective and my procedure.

Objective:      To be able to simplify expressions involving exponents. By "simplify" we mean to write the expression without using parentheses or fractional, negative, or zero exponents.

My procedure:      The six steps are listed on page 99. There is nothing sacred about the order of the steps. As stated before, I always do them in the same order so that you can follow what I am doing in the solutions.

The following example illustrates this procedure one more time.

## EXAMPLE 9

Simplify: $\left(\dfrac{\sqrt{x}}{x^2}\right)^{-2}$

Solution: $\left(\dfrac{\sqrt{x}}{x^2}\right)^{-2}$

Step 1.    $\left(\dfrac{x^{1/2}}{x^2}\right)^{-2}$

Step 2.    $\dfrac{x^{-1}}{x^{-4}}$

Step 3.    $\dfrac{x^4}{x}$

Step 4.    $x^3$

If you have made it this far, congratulations. In my opinion, you have just completed the most difficult unit in the book.

You should now be able to simplify most expressions involving exponents, whether they are fractional, negative, zero, or integers, as well as being able to simplify expressions with radicals.

Before beginning the next unit you should simplify the expressions in the exercises. The final answers should be written without fractional, negative, or zero exponents. Also, fractions should be in lowest terms. When you have completed the exercises, check your answers against those given at the back of the book. Because these problems are difficult, the complete solutions are provided for all 15.

## EXERCISES

Simplify:

1. $\dfrac{y^{2/3}}{y^{1/3}}$

2. $\left(y^{3/5}\right)^{1/4}$

3. $x^{1/2} \cdot x^{2/5}$

4. $\left(\dfrac{a^4}{c^2}\right)^{1/2}$

5. $\left[\left(\sqrt{4}\right)^{-1}\right]^2$

6. $\left(\sqrt{x}\right)^{1/2}$

7. $\left(8x^2\right)^{1/3}$

8. $\left(\dfrac{2^{-3} \cdot 2^5}{2^{-2}}\right)^3$

9. $\left(\dfrac{x^{1/3}}{x^{2/3}}\right)^3$

10. $x^{1/2} \cdot x^{5/2}$

11. $\left(\sqrt[3]{x^2}\right)^{1/2}$

12. $\dfrac{x^{-7/2} \cdot x^{3/2}}{\sqrt{x} \cdot x^{-3/2}}$

13. $\left(8\sqrt{x}\right)^{-2/3}$

14. $\left(\dfrac{27^{5/3} \cdot 27^{-1/3}}{27^{1/3}}\right)^2$

15. $\left(\dfrac{3x^{-1}}{\sqrt{x}}\right)^2$

----------------------------------------------

# UNIT 15

## Additional Practice with Exponents

Before we leave our study of exponents, we will look at a source of common errors in working with them. These are errors in dealing with the addition and subtraction of terms with exponents.

The five laws of exponents deal only with simplifying products and quotients. In fact, addition and subtraction involving terms with exponents must be simplified using the definitions covered in Units 10 and 11 on fractions.

## EXAMPLE 1

Simplify: $(ab)^{-1}$

Solution: $(ab)^{-1} = a^{-1}b^{-1} = \dfrac{1}{ab}$

## EXAMPLE 2

Simplify: $(a + b)^{-1}$.

Solution: $(a+b)^{-1} = \dfrac{1}{(a+b)} = \dfrac{1}{a+b}$

## EXAMPLE 3

Simplify: $a^{-1} + b^{-1}$.

Solution: $a^{-1} + b^{-1} = \dfrac{1}{a} \searrow \dfrac{1}{b} = \dfrac{b+a}{ab}$

The arrow is a reminder to add by using the definition from Unit 10.

Be sure you see the difference between Example 1, in which the exponent applies to the product of $a$ and $b$; Example 2, in which the exponent applies to the sum of $a$ and $b$; and Example 3, in which the exponents apply to the two separate terms.

## EXAMPLE 4

Simplify: $a + b^{-1}$.

Solution: $a + b^{-1} = a + \dfrac{1}{b} = \dfrac{a}{1} \searrow \dfrac{1}{b} = \dfrac{ab+1}{b}$

Try this problem.

## Problem 1

Simplify: $3^{-1} + x^{-1}$.

Solution:

Answer: $\dfrac{x+3}{3x}$

By now a problem like the following should be easy for you.

---

## Problem 2

Simplify:   $\dfrac{a^{-1}b^{-1}}{c^{-1}d^{-1}}$.

---

Solution:

Answer:   $\dfrac{cd}{ab}$

---

Here's another example.

## EXAMPLE 5

Simplify:   $\dfrac{a^{-1}+b^{-1}}{c^{-1}+d^{-1}}$.

---

Solution: $\dfrac{a^{-1}+b^{-1}}{c^{-1}+d^{-1}} = \dfrac{\dfrac{1}{a}+\dfrac{1}{b}}{\dfrac{1}{c}+\dfrac{1}{d}}$, which is a complex fraction

$$= \left(\frac{1}{a}+\frac{1}{b}\right) \div \left(\frac{1}{c}+\frac{1}{d}\right)$$

$$= \frac{b+a}{ab} \div \frac{d+c}{cd}$$

$$= \frac{a+b}{ab} \cdot \frac{cd}{c+d}$$

$$= \frac{cd(a+b)}{ab(c+d)}$$

---

We're going to reverse the procedure this time. I'll let you try an easier problem.

---

## Problem 3

Simplify:   $\dfrac{1}{a^{-1}+b^{-1}}$.

---

Solution:

Answer:   $\dfrac{ab}{a+b}$

---

Before proceeding, be sure you understand how Problem 2, Example 5, and Problem 3 differ.

Now consider Examples 6 and 7.

## EXAMPLE 6

Simplify:   $\dfrac{1}{x^{-1}+x}$.

---

Solution:   $\dfrac{1}{x^{-1}+x} = \dfrac{1}{\dfrac{1}{x}+x}$ , which is a complex fraction

$$= 1 \div \left(\dfrac{1}{x}+x\right)$$

$$= 1 \div \left(\dfrac{1}{x} + \dfrac{x}{1}\right)$$

$$= 1 \div \dfrac{1+x^2}{x}$$

$$= 1 \cdot \dfrac{x}{1+x^2}$$

$$= \dfrac{x}{1+x^2}$$

---

Let me repeat: the laws of exponents deal only with multiplication and division. There are no laws dealing with powers of sums and differences!

## EXAMPLE 7

Simplify:   $(x^2+a)^{-3}$

---

Solution:   $(x^2+a)^{-3} = \dfrac{1}{(x^2+a)^3}$   and that is as much as you can do!

---

You should now be able to simplify expressions in which terms with exponents are added and subtracted. Remember that, since the laws of exponents deal only with multiplication and division, you must call upon the definitions for addition and subtraction of fractions and the techniques for simplifying complex fractions.

Before beginning the next unit, simplify the expressions in the exercises, writing them as either monomials or fractions in lowest terms, with positive exponents only. None of the denominators is zero.

## EXERCISES

Simplify:

1. $(x^2 + 1)^{-2}$

2. $x^{-1} + 2^{-2}$

3. $x^{-1} - 1$

4. $\dfrac{1}{x^{-1} + y^{-1}}$

5. $3^{-1} + 3^{-2}$

6. $x^{-1} + 2y^{-1}$

7. $a + b^{-1}$

8. $5(x + y)^{-1}$

9. $\dfrac{x^{-1}}{(2x - 3)^{-2}}$

10. $3x^{-2} + y$

11. $\dfrac{a - a^{-1}}{a + a^{-1}}$   Be careful!

12. $\dfrac{3^{-1} + 2^{-1}}{3^{-1} - 2^{-1}}$

13. $\dfrac{a + b^{-1}}{ab}$

14. $\dfrac{3ab}{a^{-1} + b}$

# UNIT 16

## Multiplication of Monomials and Polynomials

In this unit we will learn how to multiply binomials and other polynomials. We will discuss some methods that should help you perform these operations and avoid common errors.

Recall that expressions with one term are called **monomials** and that expressions with more than one term (symbols or groups of symbols separated by a plus or minus sign) are called **polynomials**. Polynomials are sometimes classified according to the number of terms; for example, a **binomial** has two terms and a **trinomial** has three terms.

### MULTIPLICATION OF TWO MONOMIALS

We will first consider multiplication by a monomial.

Rule: **To multiply two monomials, multiply their numerical coefficients and find the product of the variable factors according to the laws of exponents.**

Recall that since we are multiplying (multiplication law of exponents), the exponents are added.

---

EXAMPLE 1  $(2x^3)(3x^5) = 2 \cdot 3 \cdot x^3 \cdot x^5 = 6x^8$

EXAMPLE 2  $(2xy^3)(7x^4y^5) = 2 \cdot 7 \cdot x \cdot x^4 \cdot y^3 \cdot y^5 = 14x^5y^8$

EXAMPLE 3  $(3\pi ab)(2\pi a^2bc^3) = 3 \cdot 2 \cdot \pi \cdot \pi \cdot a \cdot a^2 \cdot b \cdot b \cdot c^3 = 6\pi^2a^3b^2c^3$

---

The above rule, combined with the distributive property $a(b + c) = ab + ac$, extends in the natural way to multiplication of a polynomial by a monomial.

**EXAMPLE 4**  $2x^3(3x^5 + 2xy^3) = 6x^8 + 4x^4y^3$

**EXAMPLE 5**  $2xy^3(3x^5 + 7x^4y + 1) = 6x^6y^3 + 14x^5y^4 + 2xy^3$

**EXAMPLE 6**  $-3a^2b(2a^4b^2 - 6a^3b) = -6a^6b^3 + 18a^5b^2$

Here are three problems for you.

**Problem 1**    $3x^3(5x^2 - 2) =$

**Problem 2**    $3xy^2(2xy + 3x^2y^3) =$

**Problem 3**    $2xyz(x + y + 2z + 3) =$

Answers:   1.  $15x^5 - 6x^3$    2.  $6x^2y^3 + 9x^3y^5$    3.  $2x^2yz + 2xy^2z + 4xyz^2 + 6xyz$

# MULTIPLICATION OF TWO POLYNOMIALS

Now let's consider multiplication of two polynomials. This is the most frequently encountered problem, and ideally you should be able to do many of these multiplications mentally.

Consider, for example, $(x + 3)(x + 4)$.

Basically, three methods are commonly used to find the product of two binomials, $(x + 3)$ $(x + 4)$.

Method 1

$$
\begin{array}{r}
x + 3 \\
x + 4 \\
\hline
x^2 + 3x \\
4x + 12 \\
\hline
x^2 + 7x + 12
\end{array}
$$

This way of multiplying polynomials is not very efficient. I generally discourage its use, primarily because it necessitates rewriting the problem.

Method 2 (FOIL)

FOIL    for      F = first   $= x \cdot x$

O = outer $= 4 \cdot x$

I  = inner $= 3 \cdot x$

L = last   $= 3 \cdot 4$

F            L = F   O    I    L

$(x + 3)(x + 4) = x^2 + 4x + 3x + 12 = x^2 + 7x + 12$

I

O

Another version of the FOIL method looks like this:

$x^2$          $-10$   F   O I   L

$(x - 2)(x + 5) = x^2 + 3x - 10$

$-2x$

$5x$

$+3x$

The advantage of the FOIL method is that for simple binomials the multiplication can be performed mentally. However, it can be used only when multiplying two binomials together, *not* when multiplying a binomial by some other polynomial.

Method 3 (distributive property)

$(x + 3)(x + 4) = x(x + 4) + 3(x + 4)$

$= x^2 + 4x + 3x + 12$

$= x^2 + 7x + 12$

This method is similar to removing parentheses and works for multiplying *any* type of polynomial. I think it is the most useful method and the one I will use in all future examples.

## EXAMPLE 7

Multiply:  $(x - 7)(x + 2)$.

_ _ _ _ _ _ _ _ _ _ _ _ _ _ _ _ _ _ _ _ _ _ _ _ _ _ _ _ _ _ _ _ _ _ _ _

Solution:   $(x - 7)(x + 2) = x(x + 2) - 7(x + 2)$

$= x^2 + 2x - 7x - 14$

$= x^2 - 5x - 14$

_ _ _ _ _ _ _ _ _ _ _ _ _ _ _ _ _ _ _ _ _ _ _ _ _ _ _ _ _ _ _ _ _ _ _ _

## Problem 4

Multiply:   $(2x + 5)(3x - 2)$, using either Method 2 or Method 3.

Solution:

Answer:   $6x^2 + 11x - 10$

After this example, it will be your turn again.

## EXAMPLE 8

Multiply:   $(x + a)(2x - b)$.

Solution:   $(x + a)(2x - b) = x(2x - b) + a(2x - b)$
$$= 2x^2 - bx + 2ax - ab$$

## Problem 5

Multiply: $(3x + 1)(x - 5)$.

Solution:

Answer:   $3x^2 - 14x - 5$

## Problem 6

Multiply: $(3x - 5)(2x - 1)$.

Solution:

Answer: $6x^2 - 13x + 5$

The next example illustrates the fact that by using Method 3 we can multiply any polynomial by a binomial.

## EXAMPLE 9

Multiply: $(x + 4)(x^3 - x^2 + 3x - 1)$.

Solution: $(x + 4)(x^3 - x^2 + 3x - 1) = x(x^3 - x^2 + 3x - 1) + 4(x^3 - x^2 + 3x - 1)$
$$= x^4 - x^3 + 3x^2 - x + 4x^3 - 4x^2 + 12x - 4$$
$$= x^4 + 3x^3 - x^2 + 11x - 4$$

Note that the answer is written in descending order of the exponents, ending with the constant term.

## EXAMPLE 10

Multiply: $(x + 2)(x^3 - 1)$.

Solution: $(x + 2)(x^3 - 1) = x^4 - x + 2x^3 - 2$
$$= x^4 + 2x^3 - x - 2$$

Before completing the unit, let's look at several more examples that illustrate a "special product"—a binomial squared.

## EXAMPLE 11

Multiply: $(x + 3)^2$.

---

Solution: $(x + 3)^2 = (x + 3)(x + 3)$
$$= x(x + 3) + 3(x + 3)$$
$$= x^2 + 3x + 3x + 9$$
$$= x^2 + 6x + 9$$

Note that the answer has the first and last terms squared but there is also **a middle term that is twice the product of the two terms** of the binomial.

---

$$(a + b)^2 = (a + b)(a + b)$$
$$= a(a + b) + b(a + b)$$
$$= a^2 + ab + ab + b^2$$
$$= a^2 + 2ab + b^2$$

This problem and Example 11 illustrate the following statement:

When a binomial is squared, the result will be the first term squared plus twice the product of the two terms plus the last term squared.

$$(a+b)^2 = a^2 + 2ab + b^2$$

first term    twice the    last term
squared    product of    squared
the terms

## EXAMPLE 12

Multiply: $(x + 7)^2$.

---

Solution: $(x + 7)^2 = x^2 + 2(7x) + 7^2$
$$= x^2 + 14x + 49$$

---

## EXAMPLE 13

Multiply: $(x - 5)^2$.

---

Solution: $(x - 5)^2 = x^2 + 2(-5x) + 25$
$$= x^2 - 10x + 25$$

---

## EXAMPLE 14

Multiply:   $x(x+4)^2$.

---

Solution:   As always, remove parentheses first.

$$x(x+4)^2 = x(x^2 + 2[4x] + 16)$$
$$= x(x^2 + 8x + 16)$$
$$= x^3 + 8x^2 + 16x$$

---

## EXAMPLE 15

Multiply:   $2x^3(x-2)^2$.

---

Solution:   $2x^3(x-2)^2 = 2x^3(x^2 + 2[-2x] + 4)$
$$= 2x^3(x^2 - 4x + 4)$$
$$= 2x^5 - 8x^4 + 8x^3$$

---

You should now be able to multiply polynomials. Again I suggest using Method 3, which is similar to removing parentheses and will work for all cases.

Before beginning the next unit you should try to solve the following problems.

## EXERCISES

In each case, perform the indicated multiplication.

1.   $2cx^2(5c^2 - c - 3x)$

2.   $(x+4)(x+5)$

3.   $(x-7)(x-2)$

4.   $(x-1)(x-5)$

5.   $(x+2)(x-3)$

6.   $(a+5)^2$

7.   $(x+2)(x-2)$

8.   $(x-1)^2$

9.   $(x-4)(x+3)$

10.   $(2x+1)(x+1)$

11.   $(2x-5)(x+4)$

12.   $(3x-2)(x+7)$

13.   $(5x+1)(x+2)$

14.   $(3x+1)(3x-1)$

15.   $(2x + 3)(4x + 1)$

16.   $(5x - 1)(x + 2)$

17.   $(2x + 3)^2$

18.   $(4x - 3)^2$

19.   $(3a + b)(2a - b)$

20.   $(x + y)(x - y)$

21.   $(3x + 1)(2x - 5)$

22.   $x(x - 4)^2$

23.   $x(x - 5)^2$

24.   $(x - 2)(x^3 - 4x^2 + 7x - 1)$

25.   $(x^2 + 1)(x^2 - 3)$

26.   $(x + 2y)(x - 3y)$

27.   $(2a - 1)(3 - a)$

28.   $(x^2 - 3x + 1)(x^3 - 2x)$

29.   $(x^2 + 5)(x - 3)$

30.   $(5a - 3b)(-2a + 6b)$

# UNIT 17

## Division of Polynomials

In this unit we will examine the procedure for dividing polynomials. Although this kind of division does not occur very often, it is occasionally necessary to perform the operation in the course of solving a problem. The process outlined here will help on such occasions.

## TWO IMPORTANT DEFINITIONS

Before studying the procedure for dividing polynomials, let's look at two definitions.

> Definition: A **polynomial** in $x$ is said to be in **standard form** if:
>
> 1. All parentheses are removed.
> 2. Like terms are combined.
> 3. The terms are arranged in order of descending powers of $x$.

> Definition: The **degree of polynomial** in $x$ is the greatest power of $x$.

## EXAMPLE 1

Write $x + 2x(x^2 - 5)$ in standard form and find its degree.

---

Solution:  $x + 2x(x^2 - 5) = x + 2x^3 - 10x$

$$= 2x^3 - 9x$$

Standard form: $2x^{\textcircled{3}} - 9x$

Degree: 3

---

# PROCEDURE FOR DIVIDING POLYNOMIALS

Now let's look at the procedure for dividing polynomials. Consider this problem:

Divide: $(5x^2 - 3x + 1)$ by $(x - 2)$.

The basic procedure parallels the long-division process in arithmetic.

Step 1: Arrange both dividend and divisor in standard form and set up as a long-division problem, leaving spaces for any missing terms.

$$x - 2 \overline{)5x^2 - 3x + 1}$$

Step 2: Divide the first term of the divisor into the first term of the dividend. (Recall that, since we are dividing, the exponents are subtracted.)

$$x - 2 \overline{)5x^2 - 3x + 1} \quad \overset{5x}{}$$

Step 3: Multiply each term of the divisor by the first term of the quotient.

$$\begin{array}{r} 5x \\ x - 2 \overline{)5x^2 - 3x + 1} \\ 5x^2 - 10x \end{array}$$

Step 4: Subtract like terms and bring down the next term from the dividend.

$$\begin{array}{r} 5x \\ x - 2 \overline{)5x^2 - 3x + 1} \\ 5x^2 - 10x \\ +7x +1 \end{array}$$

Step 5: Repeat steps 2, 3, and 4, using the new remainder, $7x + 1$, as the dividend.

$$\begin{array}{r} 5x + 7 \\ x - 2 \overline{)5x^2 - 3x + 1} \\ 5x^2 - 10x \\ 7x +1 \end{array}$$

Step 6: Continue repeating steps 2, 3, and 4 until the degree of the remainder is less than the degree of the divisor.

$$\begin{array}{r} 7x - 14 \\ \hline 15 \end{array}$$

Therefore

$$(5x^2 - 3x + 1) \text{ divided by } (x - 2) \text{ equals } 5x + 7 + \frac{15}{x - 2}.$$

Note that the remainder is written as a fraction.

Let me now try condensing some of the explanation.

## EXAMPLE 2

Divide:   $(5x^2 + 7x - 1)$ by $(x + 3)$.

Solution:   Step 2.   Divide first terms:

$$\frac{5x^2}{x} = 5x.$$

Step 3.   Multiply:

$$5x(x + 3) = 5x^2 + 15x.$$

Step 4.   Subtract; bring down $- 1$.

Repeat steps 2, 3, and 4, using $-8x - 1$.

Step 2.   Divide first terms:

$$\frac{-8x}{x} = -8.$$

Step 3.   Multiply:

$$-8(x + 3) = -8x - 24.$$

Step 4.   Subtract.

$$x+3 \overline{)5x^2 + 7x - 1} \quad \begin{array}{l} 5x \\ \\ 5x^2 + 15x \\ -8x - 1 \end{array}$$

$$x+3 \overline{)5x^2 + 7x - 1} \quad \begin{array}{l} 5x - 8 \\ \\ 5x^2 + 15x \\ -8x - 1 \\ -8x - 24 \\ 23 \end{array}$$

The division process is finished because the degree of the remainder is less than the degree of the divisor.

Write the remainder as a fraction.

$$x+3 \overline{)5x^2 + 7x - 1} \quad \begin{array}{l} 5x - 8 + \dfrac{23}{x+3} \\ \\ 5x^2 + 15x \\ -8x - 1 \\ -8x - 24 \\ +23 \end{array}$$

Therefore:

$$(5x^2 + 7x - 1) \div (x + 3) = 5x - 8 + \frac{23}{x+3}.$$

Here are two more examples. Example 3 is given in detail, with explanation.

## EXAMPLE 3

Divide: $(6x^2 + 23x + 20)$ by $(2x + 5)$

Solution: Step 1. Set up problem.

Step 2. Divide first terms:

$$\frac{6x^2}{2x} = 3x.$$

Step 3. Multiply:

$$3x(2x + 5).$$

Step 4. Subtract and bring down

$$
\begin{array}{r}
3x \;\; +4 \\
2x+5{\overline{\smash{\big)}\,6x^2+23x+20}} \\
\underline{6x^2+15x} \phantom{0000} \\
8x+20 \\
\underline{8x+20} \\
0
\end{array}
$$

the next term of 20.

Repeat steps 2, 3, and 4 using $8x + 20$.

Step 2. Divide first terms:

$$\frac{8x}{2x} = 4.$$

Step 3. Multiply:

$$4(2x + 5).$$

Step 4. Subtract.

The division process is finished, and there is no remainder.

Answer: $3x + 4$

## EXAMPLE 4

Divide: $(x^2 - x - 15)$ by $(x + 1)$

Solution:

$$
\begin{array}{r}
x - 2 \phantom{xxxx} \\
x + 1 \overline{)\, x^2 - x - 15} \\
\underline{x^2 + x} \quad \downarrow \\
-2x - 15 \\
\underline{-2x - 2} \\
-17
\end{array}
$$

Answer: $\quad x - 2 + \dfrac{-17}{x+1}$ .

- - - - - - - - - - - - - - - - - - - - - - - - - - - - - - - - - - - -

You try this one.

─────────────────────────────────────────────

## Problem 1

Divide:  $(3x^2 + 4x + 1)$ by $(3x + 1)$.

- - - - - - - - - - - - - - - - - - - - - - - - - - - - - - - - - - - -

Solution:

Answer:  $x + 1$

- - - - - - - - - - - - - - - - - - - - - - - - - - - - - - - - - - - -

Look carefully at Examples 5–7.

═══════════════════════════════════════════════

## EXAMPLE 5

Divide:  $(6x + 5)$ by $(2x - 1)$

- - - - - - - - - - - - - - - - - - - - - - - - - - - - - - - - - - - -

Solution:

$$
\begin{array}{r}
3 \phantom{xxx} \\
2x - 1 \overline{)\, 6x + 5} \\
\underline{6x - 3} \\
+8
\end{array}
$$

Answer:  $3 + \dfrac{8}{2x - 1}$

- - - - - - - - - - - - - - - - - - - - - - - - - - - - - - - - - - - -

## EXAMPLE 6

Divide: $(4x^2 - 7)$ by $(3 + x)$.

Solution:

$$
\begin{array}{r}
4x - 12 \phantom{000000} \\
x+3{\overline{\smash{\big)}\,4x^2 + \phantom{0}0 - 7}} \\
\underline{4x^2 + 12x} \phantom{0000} \\
-12x - 7 \phantom{00} \\
\underline{-12x - 36} \\
29
\end{array}
$$

Answer: $4x - 12 + \dfrac{29}{x+3}$

Note: The divisor had to be put in standard form, and a space had to be left for the missing $x$-term in the dividend. I inserted a zero in the space to avoid confusion when I subtracted.

## EXAMPLE 7

Divide: $(x^3 - 8)$ by $(x - 2)$

Solution:

$$
\begin{array}{r}
x^2 + 2x + 4 \phantom{0000} \\
x-2{\overline{\smash{\big)}\,x^3 + 0 \phantom{0} + 0 \phantom{0} - 8}} \\
\underline{x^3 - 2x^2} \phantom{0000000} \\
+2x^2 + 0 \phantom{0000} \\
\underline{+2x^2 - 4x} \phantom{00} \\
+4x - 8 \\
\underline{+4x - 8} \\
0
\end{array}
$$

Answer: $x^2 + 2x + 4$

You should now be able to divide a polynomial by a binomial. Remember that the procedure is identical to the long-division process in arithmetic.

Recall that, in short, the division procedure is:

1. **Set up** as long division.
2. **Divide** first terms.
3. **Multiply** the quotient times the divisor.
4. **Subtract** like terms; **bring down** the next term.
5. **Repeat**, using the new remainder as the dividend.
6. **Continue** until the degree of the remainder is less than the degree of the divisor.

Now try the following problems.

## EXERCISES

Divide:

1. $(x^2 + 5x + 6) \div (x + 2)$
2. $(12x + 1) \div (x - 4)$
3. $(x^2 + 8x + 16) \div (x + 5)$
4. $(10x^2 - 13x + 1) \div (5x + 1)$
5. $(3x^2 - 10x - 8) \div (3x + 2)$
6. $(6x^2 - 19x + 10) \div (2x - 1)$
7. $(4x^2 - 49) \div (2x - 7)$
8. $(5x^2 - 1) \div (x - 1)$
9. $(x^3 + 1) \div (x + 1)$
10. $(x^3 + 2) \div (x + 2)$

# UNIT 18

## Factoring Polynomials

In this unit we will discuss **factoring** of polynomials. Simply stated, factoring involves finding factors whose products are equal to the original polynomial. We will look first at the simplest case, that of finding a common monomial factor. Then we will consider the factoring of trinomials.

### FACTORING OUT A COMMON MONOMIAL FACTOR

First, recall the distributive law:   $a(b + c) = ab + ac$.

$$\text{multiplication} \longrightarrow \quad \longleftarrow \text{factoring}$$

Factoring out a common monomial factor is the application of the distributive law read from right to left.

A common monomial factor is a single expression that is a factor of each term of the polynomial.

Factoring out the common monomial factor is, in effect, taking out in front of the parentheses, *as much as possible*, what is common to all terms of the polynomial. It is just the reverse of removing parentheses. You already did this on a limited basis when you solved literal equations in Unit 5.

The test of factoring is always whether you can multiply the factors together and obtain the original polynomial.

| EXAMPLE 1 | $6x^2y + 5x^2z = x^2(6y + 5z)$ |
| EXAMPLE 2 | $2ay + 2 = 2(ay + 1)$ |
| EXAMPLE 3 | $2xy - xy^2 + 3x^2y = xy(2 - y + 3x)$ |
| EXAMPLE 4 | $8x^2 - 12x = 4x(2x - 3)$ |
| EXAMPLE 5 | $12a^3 - 18a^2 + 3a = 3a(4a^2 - 6a + 1)$ |

Whenever you are factoring an expression, always first look to see whether there is a common monomial factor! Often the common monomial factor is overlooked, and the resulting problem is more difficult to handle.

# FACTORING TRINOMIALS

Let us first consider factoring trinomials that are polynomials of degree 2 with the coefficient of $x^2$ being 1. In this unit such trinomials are factored experimentally—by trial and error—if they are factorable at all. However, the number of trials can often be reduced if certain observations are made.

We will classify trinomials into three categories, Cases I, II, and III, as follows.

Let $D$, $E$, $a$, and $b$ be whole numbers.

In words, Case I is a trinomial with all plus signs. If the trinomial can be factored, it will factor into two binomials of the form $(x + a)(x + b)$, where $a$ and $b$ are whole numbers such that their product $ab$ equals the constant of the trinomial $E$ and their sum $a + b$ equals the coefficient of the $x$ term of the trinomial $D$. The approach is based on applying the FOIL method in reverse.

A word of caution: Factoring can be presented in various manners. Some authors, professors, and teachers explain factoring using integers, whereas others prefer to use whole numbers. Obviously both sets of rules result in the same final answer. In this book, factoring will be explained in terms of whole numbers. Thus, we will **always** be looking for whole numbers, not integers and not fractions, to satisfy the various conditions.

A second word of caution: Not all polynomials can be factored. If no whole numbers exist to satisfy the necessary conditions, the polynomial is said to be **prime**.

## EXAMPLE 6

Factor:   $x^2 + 8x + 15$.

Solution:   $x^2 + 8x + 15 = (x + ?)(x + ?)$

$x^2 + Dx + E = (x + a)(x + b)$   where the product   $ab = E = 15$

and the sum   $a + b = D = 8$

So we are looking for two whole numbers whose product is 15 and whose sum is 8. The numbers are 3 and 5.

Therefore:

$$x^2 + 8x + 15 = (x + 3)(x + 5)$$

## EXAMPLE 7

Factor:   $x^2 + 13x + 12$.

Solution:   $x^2 + 13x + 12 = (x + ?)(x + ?)$

$x^2 + Dx + E = (x + a)(x + b)$   where the product   $ab = E = 12$

and the sum   $a + b = D = 13$

We need two whole numbers whose product is 12 and whose sum is 13. The numbers are 12 and 1.

Therefore:

$$x^2 + 13x + 12 = (x + 12)(x + 1)$$

## EXAMPLE 8

Factor:   $x^2 + 7x + 12$.

Solution:   We need two whole numbers whose product is 12 and whose sum is 7. The numbers are 3 and 4.

Therefore:

$$x^2 + 7x + 12 = (x + 3)(x + 4)$$

The following are more example of Case I. You might cover the answers and try factoring them yourself.

EXAMPLE  9     $x^2 + 10x + 9 = (x + 9)(x + 1)$

EXAMPLE 10     $x^2 + 5x + 6 = (x + 2)(x + 3)$

EXAMPLE 11     $x^2 + 13x + 22 = (x + 2)(x + 11)$

EXAMPLE 12     $x^2 + 8x + 12 = (x + 2)(x + 6)$

EXAMPLE 13     $x^2 + 12x + 35 = (x + 5)(x + 7)$

## EXAMPLE 14

Factor:     $x^2 + 7x + 1$

Solution:   We are looking for two whole numbers whose product is 1 and whose sum is 7. No such whole numbers exist.

Therefore:

$$x^2 + 7x + 1 \text{ is prime, that is, not factorable}$$

---

Case II.

$x^2 - Dx + E = (x - a)(x - b)$   where the product   $ab = E$
                                  and the sum        $a + b = D$

↑

middle term only negative

---

Case II is a trinomial in which *only* the middle term is negative. If a trinomial can be factored, it will factor into two binomials of the form $(x - a)(x - b)$, where, as in Case I, $a$ and $b$ are whole numbers such that their product equals $E$ and their sum equals $D$.

## EXAMPLE 15

Factor:   $x^2 - 9x + 14$.

Solution:   $x^2 - 9x + 14 = (x - ?)(x - ?)$

$x^2 - Dx + E = (x - a)(x - b)$   where the product   $ab = E = 14$

and the sum        $a + b = D = 9$

We are looking for two whole numbers whose product is 14 and whose sum is 9. The numbers are 7 and 2.

Therefore:

$$x^2 - 9x + 14 = (x - 2)(x - 7)$$

## EXAMPLE 16

Factor:    $x^2 - 7x + 12$.

_____

Solution:    $x^2 - 7x + 12 = (x - ?)(x - ?)$

$x^2 - Dx + E = (x - a)(x - b)$    where the product   $ab = E = 12$

and the sum        $a + b = D = 7$

The numbers are 3 and 4.

Therefore:

$$x^2 - 7x + 12 = (x - 3)(x - 4)$$

_____

Case III.

$x^2 \pm Dx - E = (x + a)(x - b)$    where the product        $ab = E$

and the difference of $a$ and $b = D$

and the larger of $a$ and $b$ has the sign of the middle term

last term negative

Case III is a trinomial with the *last* term negative. If a trinomial can be factored, it will factor into the form $(x + a)(x - b)$, where $a$ and $b$ are whole numbers such that, as before, their product equals $E$. But in this case the difference of $a$ and $b$ must equal $D$, and the larger of the two numbers has the same sign as the middle term of the trinomial.

I realize all that is a bit wordy. I hope Examples 17 and 18 will clarify any questions you may have.

## EXAMPLE 17

Factor:    $x^2 - 3x - 10$.

_____

Solution:    $x^2 - 3x - 10 = (x + ?)(x - ?)$

$x^2 - Dx - E = (x + a)(x - b)$    where the product        $ab = E = 10$

and the difference of $a$ and $b = D = 3$

We are looking for two whole numbers whose product is 10 and whose difference is 3. Obviously, the numbers are 2 and 5.

Since the middle term is negative, the 5 goes with the negative sign. Therefore:

$$x^2 - 3x - 10 = (x + 2)(x - 5)$$

Remember:   the test of factoring is always whether you can multiply and obtain the original polynomial.

$$x^2 \qquad - 10$$

Using the FOIL method:   $(x + 2)(x - 5) = x^2 - 3x - 10$

and our factors
were correct.

$$2x$$
$$-5x$$

$$-3x$$

---

# EXAMPLE 18

Factor:   $x^2 + x - 20$.

---

Solution:   $x^2 + x - 20 = (x + ?)(x - ?)$

$x^2 + Dx - E = (x + a)(x - b)$   where the product $\qquad ab = 20$

and the difference of $a$ and $b = D = 1$

The numbers are 4 and 5. Since the middle term is positive, the 5 goes with the plus. Therefore:

$$x^2 + x - 20 = (x + 5)(x - 4)$$

---

The following seven problems are more examples of Case III for practice. Try factoring the trinomials yourself.

---

Problem 1     $x^2 - 4x - 21$

Problem 2     $x^2 + 5x - 6$

Problem 3     $x^2 + 6x - 16$

Problem 4     $x^2 - x - 56$

Problem 5     $x^2 + 2x - 15$

Problem 6     $x^2 + 8x - 9$

Problem 7     $x^2 - x - 7$

---

Answers:   1. $(x + 3)(x - 7)$     2. $(x + 6)(x - 1)$     3. $(x + 8)(x - 2)$
4. $(x + 7)(x - 8)$     5. $(x + 5)(x - 3)$     6. $(x + 9)(x - 1)$     7. Prime

---

If we let $D$, $E$, $a$, and $b$ be integers, a generalized case can be used to represent all three of the trinomials illustrated in Cases I, II, and III as shown in the accompanying box. As stated at the beginning of this unit, I prefer to look at three separate cases using whole numbers, but the choice is always yours if you prefer to use integers.

---

Generalized Case. $x^2 + Dx + E = (x + a)(x + b)$ where E is the product of $a$ and $b$,
D is the **algebraic** sum of $a$ and $b$, with D, E, a, and b integers

---

If we let $D$, $E$, $a$, and $b$ be integers, the generalized case represents all three of the trinomials illustrated in Cases I, II, and III. This trinomial will factor into two binomials of the form $(x + a)(x + b)$, where $a$ and $b$ are integers such that their product equals $E$ and their **algebraic** sum equals $D$.

If $E$ is negative, $a$ and $b$ must have opposite signs because only the product of unlike signs yields a negative number.

If $E$ is positive, $a$ and $b$ must have the same sign, either both positive or both negative.

Since $D$ is the algebraic sum, its sign determines the sign of the numerically larger of $a$ and $b$ (whether their signs are the same or different).

Before beginning the next unit, try factoring the following polynomials. Remember to always first look to see whether there is a common monomial factor that can be factored out before attempting to factor the polynomial into binomial factors.

===

## EXERCISES

Factor

1. $x^2 + 2x - 3$
2. $x^2 - 15x + 56$
3. $x^2 + x$
4. $3x^2y - 12xy^2$
5. $3bx^2 + 27b^2$
6. $x^2 - x - 6$
7. $x^2 + 5x + 6$
8. $x^2 - 7x + 12$
9. $x^2 - x + 5$
10. $x^2 - 2x - 8$
11. $x^2 + 5x + 3$
12. $2x^3 + 2x^2 + 22x$

13. $5x^2 - 5x - 5$

14. $x^2 + 3x - 10$

15. $x^2 + 7x - 30$

16. $x^2 + 7x + 6$

17. $x^2 + 2x + 1$

18. $x^2 - 6x + 9$

19. $x^2 - x - 56$

20. $x^2 + 4x - 45$

21. $x^2 + 16x + 64$

22. $x^2 - 13x + 40$

23. $x^2 + 7x - 18$

24. $x^3 + x^2 + 5x$

25. $x^2 - 10x + 21$

26. $x^2 - 7x - 18$

# UNIT 19

## Factoring a Special Binomial

In this unit we introduce a frequently occurring formula that will aid you in factoring. Also, we consider a suggested procedure to follow when factoring any polynomial.

## AN IMPORTANT FORMULA

It is helpful to memorize the formula for factoring the difference of two squares. This will help you recognize when it is needed.

$$\text{Difference of two squares:} \quad x^2 - y^2 = (x + y)(x - y)$$

The difference of two squares will factor into two binomials of the form $(x + y)(x - y)$, where $x$ is the square root of the first term and $y$ is the square root of the second term.

### EXAMPLE 1

Factor: $a^2 - 25$.

----

Solution: The square root of $a^2$ is $a$.
The square root of 25 is 5.
Therefore:

$$a^2 - 25 = (a + 5)(a - 5)$$

The difference of two squares factors to the sum of their square roots times the difference of their square roots.

----

| | |
|---|---|
| EXAMPLE 2 | $x^2 - 1 = (x + 1)(x - 1)$ |
| EXAMPLE 3 | $9x^2 - 4y^2 = (3x + 2y)(3x - 2y)$ |
| EXAMPLE 4 | $16a^2 - 49 = (4a + 7)(4a - 7)$ |
| EXAMPLE 5 | $x^2y^2 - 100 = (xy + 10)(xy - 10)$ |
| EXAMPLE 6 | $y^2 + 25$ Prime—not factorable. This is the *sum* of two squares. |
| EXAMPLE 7 | $x^3 - 4$ Prime—$x^3$ is *not* a square. |

More often than not, to factor a polynomial completely requires more than one step. The following is a suggested procedure for factoring.

## SUGGESTED PROCEDURE FOR FACTORING

1. First factor out any common monomial factor.
2. If there are two terms, check for the difference of two squares and factor accordingly.
3. If there are three terms, decide whether the trinomial is Case I, II, or III and try factoring the trinomial as the products of two binomials.

### EXAMPLE 8

Factor completely: $3x^4 - 27x^2$.

Solution:

common monomial factor
difference of two squares; factor again

$$3x^4 - 27x^2 = 3x^2(x^2 - 9)$$
$$= 3x^2(x + 3)(x - 3)$$

### EXAMPLE 9

Factor completely: $3x^4y - 6x^3y + 3x^2y$.

Solution:

common monomial factor
three terms—Case II; factor again

$$3x^4y - 6x^3y + 3x^2y = 3x^2y(x^2 - 2x + 1)$$
$$= 3x^2y(x - 1)(x - 1)$$

## EXAMPLE 10

Factor completely:    $x^4 - 16$.

---

Solution:   This is the difference of two squares.

The square root of $x^4$ is $x^2$.

The square root of 16 is 4.

Therefore:

difference of two squares;
factor again
$$x^4 - 16 = (x^2 + 4)(x^2 - 4)$$
$$= (x^2 + 4)(x + 2)(x - 2)$$

---

## EXAMPLE 11

Factor completely:    $2x^3 + 8x^2 + 6x$.

---

Solution:

common monomial factor

three terms—Case I;
factor again

$$2x^3 + 8x^2 + 6x = 2x(x^2 + 4x + 3)$$
$$= 2x(x + 1)(x + 3)$$

---

## EXAMPLE 12

Factor:    $(x + 3)^2 - 16$.

---

Solution:   This is the difference of two squares.

The square root of $(x + 3)^2$ is $x + 3$.

The square root of 16 is 4.

Therefore:

$$(x + 3)^2 - 16 = [(x + 3) + 4][(x + 3) - 4]$$
$$= (x + 7)(x - 1)$$

---

You should now have a good understanding of factoring and be able to factor many expressions. Remember our suggested three-step procedure which, simply stated, is as follows:

1. Factor out any common monomial factor.

2. If there are two terms, check for difference of two squares and factor accordingly.

3.  If there are three terms, factor into two binomials, using Case I, II or III. Before beginning the next unit, try factoring the following polynomials.

## EXERCISES

Factor completely:

1.  $x^2 - 36$

2.  $w^2 - 81$

3.  $x^2 - y^2$

4.  $16x^2 - 9$

5.  $3b^2 - 75$

6.  $9 - 6x + x^2$

7.  $2x^2 - 2$

8.  $3abc^2 - 3abd^2$

9.  $x^2 + 2x - 8$

10. $x^2 - x + 7$

11. $x^3 - 36x$

12. $x^2 + 4$

13. $x^2 + 13x + 30$

14. $3r^3 - 6r^2 - 45r$

15. $x^2 + 5x - 14$

16. $2a^2b^2c^2 - 4ab^2c^2 + 2b^2c^2$

17. $5x^2y - 15xy - 10y$

18. $5x^4 + 10x^3 - 15x^2$

19. $3x^2 - 12$

20. $a^2b^2 - a^2c^2$

21. $2xy^2 - 54xy + 100x$

22. $10ab^2 - 140ab + 330a$

23. $w^2x^2y^2 + 7w^2x^2y - 18w^2x^2$

24. $2ax^2 - 2ax - 40a$

25. $4a^2 - 9b^2$

26. $2x^2 - 10x - 12$

27. $4r^3s^2 - 48r^2s^2 + 108rs^2$

28. $2y^2z + 38yz + 96z$

29. $3a^2b^5 - 3a^2b$

30. $a^2x^4 - 81a^2$

# UNIT 20

## Factoring (Continued)

This unit will continue our discussion of factoring. Specifically, we will learn a technique called **factoring by grouping**. Additionally, we will learn to factor trinomials in which the coefficient of $x^2$ is not 1. The unit concludes with a general strategy for factoring.

## FACTORING BY GROUPING

Thus far we have looked at procedures for factoring polynomials with two or three terms. We will now examine a technique for factoring a polynomial with four terms. It is called **factoring by grouping**. As the name suggests, we try grouping the four terms into pairs that have some common variable. Notice I use the word *try*. Not all polynomials with four terms can be factored using this technique. Several examples should be sufficient.

### EXAMPLE 1

Factor: $ab + ac - bd - cd$.

---

Solution:  Terms one and two have $a$ in common.

Terms three and four have $d$ in common and are both negative.
Therefore:

$$ab + ac - bd - cd = a(b + c) - d(b + c)$$

Since we can treat $(b + c)$ as a single quantity and $(b + c)$ is common to both terms, factor it out as a common factor:

$$= (b + c)(a - d)$$

Neat technique, isn't it?

---

## EXAMPLE 2

Factor: $2c - 6 - cy + 3y$.

Solution: $2c - 6 - cy + 3y = 2(c - 3) - y(c - 3)$

Since $(c - 3)$ is common to both terms, factor it out in front.

$$= (c - 3)(2 - y)$$

## EXAMPLE 3

Factor: $2ax - 4bx + ay - 2by$.

Solution: $2ax - 4bx + ay - 2by = 2x(a - 2b) + y(a - 2b)$

$(a - 2b)$ is common; factor it out in front.

$$= (a - 2b)(2x + y)$$

## EXAMPLE 4

Factor: $12x^2 - 9x + 8x - 6$.

Solution: $12x^2 - 9x + 8x - 6 = 3x(4x - 3) + 2(4x - 3)$

$(4x - 3)$ is common; factor it out in front.

$$= (4x - 3)(3x + 2)$$

It's your turn to try a problem.

## Problem 1

Factor: $2axy + 4ay - 3x - 6$.

Solution:

Answer: $(x + 2)(2ay - 3)$

---

# FACTORING TRINOMIALS IN WHICH THE COEFFICIENT OF $x^2$ IS NOT 1

When the coefficient of $x^2$ is not 1, the problem of factoring trinomials can become far more complicated. Two procedures are presented here, both of which are similar to that of Unit 18. The first one involves some trial and error, whereas the second uses factoring by grouping.

## EXAMPLE 5

Factor: $3x^2 + 4x + 1$.

---

Solution: Since there are only plus signs, the trinomial is similar to Case I and the two binomials will both have plus signs.

$$3x^2 + 4x + 1 = (\ \ + \ \ )(\ \ + \ \ )$$

Recall from the FOIL method diagram that:
the product of the first terms must be $3x^2$,
the product of the last terms must be 1, and
the sum of inner product and outer product must be $4x$.

To get $3x^2$ the factors are $3x$ and $x$.
To get 1 the factors are 1 and 1.
That leaves us with the possible binomial factors being $(3x + 1)(x + 1)$.

Checking to see whether the middle term is correct:

$$
\begin{array}{r}
x \\
+3x \\
\hline
4x
\end{array}
$$

Hence:

$$3x^2 + 4x + 1 = (3x + 1)(x + 1)$$

Remember that we can always verify our answer by multiplying; we should obtain the original polynomial.

---

## EXAMPLE 6

Factor:   $2x^2 - 5x + 3$.

---

Solution:   Since the middle term only is negative, the trinomial is similar to Case II and the binomial factors will both have negative signs.

$$2x^2 - 5x + 3 = (\ -\ )(\ -\ )$$

To get $2x^2$ the factors are $2x$ and $x$.
To get 3 the factors are 1 and 3.
That leaves us with several choices.
The possible binomial factors are either

$$(2x - 1)(x - 3) \qquad \text{or} \qquad (2x - 3)(x - 1)$$

$$\begin{array}{c} -x \\ -6x \\ \hline -7x \end{array} \qquad\qquad \begin{array}{c} -3x \\ -2x \\ \hline -5x \end{array}$$

We try each one to see which has the correct middle term. Therefore:

$$2x^2 - 5x + 3 = (2x - 3)(x - 1)$$

---

## EXAMPLE 7

Factor:   $6x^2 - x - 15$.

---

Solution:   Since the last term is negative, the trinomial is similar to Case III and the binomial factors will have different signs.

$$6x^2 - x - 15 = (\ +\ )(\ -\ )$$

To get $6x^2$ the possible factors are:

$$(6x\quad)(x\quad) \qquad \text{or} \qquad (2x\quad)(3x\quad)$$

Possible factors for 15 are $15 \cdot 1$ and $5 \cdot 3$. So the possibilities are:

| | | |
|---|---|---|
| $(6x + 1)(x - 15)$ | $(6x - 1)(x + 15)$ | $(3x + 3)(2x - 5)$ |
| $(6x + 15)(x - 1)$ | $(6x - 15)(x + 1)$ | $(3x - 3)(2x + 5)$ |
| $(6x + 3)(x - 5)$ | $(6x - 3)(x + 5)$ | $(3x + 5)(2x - 3)$ |
| $(6x + 5)(x - 3)$ | $(6x - 5)(x + 3)$ | $(3x - 5)(2x + 3)$ |
| $(3x + 1)(2x - 15)$ | $(3x - 1)(2x + 15)$ | |
| $(3x + 15)(2x - 1)$ | $(3x - 15)(2x + 1)$ | |

Fortunately, eight of the possibilities can be eliminated simply by observing that, since there is no common factor in the original trinomial, there can be none in any of its factored forms. For example, $(6x + 15)$ is impossible as a factor since $(6x + 15) = 3(2x + 5)$, but the original trinomial does not have a common factor of 3.

With this observation the list of binomial factors to be considered can be shortened to:

$(6x + 1)(\ x - 15)$ $\qquad$ $(6x - 1)(\ x + 15)$ $\qquad$ $(3x + 5)(2x - 3)$
$(6x + 5)(\ x - \ 3)$ $\qquad$ $(6x - 5)(\ x + \ 3)$ $\qquad$ $(3x - 5)(2x + 3)$
$(3x + 1)(2x - 15)$ $\qquad$ $(3x - 1)(2x + 15)$

Upon inspection we find that the binomial factors are:

$$(3x - 5)(2x + 3)$$

$$-10x$$
$$9x$$
$$-x$$

Therefore:

$$6x^2 - x - 15 = (3x - 5)(2x + 3)$$

---

This required quite a bit of writing and is not to my liking. I prefer a different procedure that does not depend on trial and error. I think you'll find it easier to use.

In Unit 18 we considered polynomials of degree 2 where the coefficient of $x^2$ was 1 and classified trinomials into three categories, Cases I, II, and III. Now we will expand each category to include polynomials where the coefficient of $x^2$ is $A$. As before, let $A$, $D$, $E$, $a$, and $b$ be whole numbers.

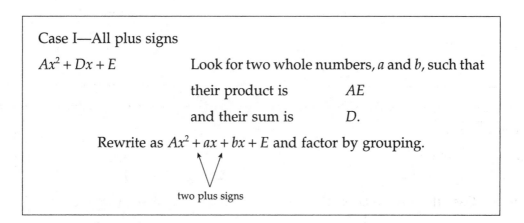

Case I—All plus signs

$Ax^2 + Dx + E$ $\qquad$ Look for two whole numbers, $a$ and $b$, such that

their product is $\qquad$ $AE$

and their sum is $\qquad$ $D$.

Rewrite as $Ax^2 + ax + bx + E$ and factor by grouping.

two plus signs

Notice that this strategy involves rewriting the linear term, $Dx$, as $ax + bx$ so that factoring by grouping can be used.

I will use the same three examples previously worked to illustrate the simplicity of this technique.

# EXAMPLE 8

Factor:   $3x^2 + 4x + 1$.

_____

Solution:   Case I:    $Ax^2 + Dx + E$

$3x^2 + 4x + 1$

product:    $AE = 3 \cdot 1 = 3$

sum:        $D = 4$

We need two whole numbers whose product is 3 and whose sum is 4. The numbers are 3 and 1. Rewrite the original trinomial using 3 and 1 as the coefficients of $x$, both with positive signs, and factor by grouping.

the two whole numbers

$$3x^2 + 4x + 1 = 3x^2 + 3x + x + 1$$
$$= 3x(x + 1) + (x + 1)$$
$$= (x + 1)(3x + 1)$$

_____

Case II—Only middle term is negative

$Ax^2 - Dx + E$          Look for two whole numbers, $a$ and $b$, such that

their product is          $AE$

and their sum is          $D$.

Rewrite as $Ax^2 - ax - bx + E$ and factor by grouping.

two minus signs

# EXAMPLE 9

Factor:   $2x^2 - 5x + 3$.

_____

Solution:   Case II:    $Ax^2 - Dx + E$

$2x^2 - 5x + 3$

product:    $AE = 2 \cdot 3 = 6$

sum:        $D = 5$

We need two whole numbers whose product is 6 and whose sum is 5. The numbers are 2 and 3. Rewrite the original trinomial using 2 and 3 as the coefficients of $x$, both with minus signs, and factor by grouping.

the two whole numbers

$$2x^2 - 5x + 3 = 2x^2 - 2x - 3x + 3$$
$$= 2x(x - 1) - 3(x - 1)$$
$$= (x - 1)(2x - 3)$$

---

Case III—Last term is negative

$Ax^2 \pm Dx - E$      Look for two whole numbers, $a$ and $b$, such that

their product is      $AE$

and their difference is    $D$.

Rewrite as $Ax^2 + ax - bx - E$, with the larger of $a$ and $b$ having the sign of the middle term, and factor by grouping.

opposite signs

---

## EXAMPLE 10

Factor:   $6x^2 - x - 15$.

---

Solution:   Case III:   $Ax^2 \pm Dx - E$

$$6x^2 - x \quad - 15$$

product:     $AE = 6 \cdot 15 = 90$

difference:    $D = 1$

We need two whole numbers whose product is 90 and whose *difference* is 1. The numbers are 9 and 10. Rewrite the original trinomial, using 9 and 10 as the coefficients of $x$, with opposite signs. Since the middle term is negative, the 10 goes with the negative sign.

opposite signs

$$6x^2 - x - 15 = 6x^2 + 9x - 10x - 15$$
$$= 3x(2x + 3) - 5(2x + 3)$$
$$= (2x + 3)(3x - 5)$$

---

I suggest you go back and look at Example 7 and compare the amount of work required for the two procedures. I think you'll see why I prefer the second method.

To conclude this unit, we can now generalize factoring into a basic strategy.

### Factoring: A General Strategy

1. **First**—factor out any common monomial factors.

2. If there are **two** terms, check for the difference of two squares and factor accordingly.

3. If there are **three** terms, decide whether the trinomial is Case I, II, or III and look for two whole numbers, $a$ and $b$, such that if given:

*Case I—All plus signs*

$x^2 + Dx + E$

| | |
|---|---|
| their product is | $E$ |
| their sum is | $D$ |

and the factors will be $(x + a)(x + b)$.

$Ax^2 + Dx + E$

| | |
|---|---|
| their product is | $AE$ |
| their sum is | $D$. |

Rewrite as $Ax^2 + ax + bx + E$ and factor by grouping.

*Case II—Only middle term is negative*

$x^2 - Dx + E$

| | |
|---|---|
| their product is | $E$ |
| their sum is | $D$ |

and the factors will be $(x - a)(x - b)$.

$Ax^2 - Dx + E$

| | |
|---|---|
| their product is | $AE$ |
| their sum is | $D$. |

Rewrite as $Ax^2 - ax - bx + E$ and factor by grouping.

*Case III—Last term is negative*

$x^2 \pm Dx - E$

| | |
|---|---|
| their product is | $E$ |
| their difference is | $D$ |

and the factors will be $(x + a)(x - b)$
and the larger number goes with the sign of the middle term.

$Ax^2 \pm Dx - E$

| | |
|---|---|
| their product is | $AE$ |
| their difference is | $D$ |

Rewrite as $Ax^2 + ax - bx - E$ and the larger number goes with the sign of the middle term and factor by grouping.

4. If there are **four** terms, try factoring by grouping.

As I stated in Unit 18, if we let $A$, $D$, $E$, $a$, and $b$ be integers, a generalized case can be used to represent all three trinomials. I prefer to look at three separate cases using whole numbers, but the choice is yours.

It's time for you to apply what you have learned in this unit.

## Problem 2

Factor:   $3x^2 + 13x + 4$.

Solution:   The trinomial is an example of Case_____. We need two whole numbers whose product is (_____)(_____) = _____ and whose sum is _____.

The numbers are _____ and _____. Therefore:

$$3x^2 + 13x + 4 = 3x^2 \underline{\hspace{4cm}} + 4$$

$$= $$

$$= $$

Answer:   $(3x + 1)(x + 4)$

## Problem 3

Factor:   $2x^2 + 7x - 15$.

Solution:

Answer:   $(2x - 3)(x + 5)$

By now you might be saying to yourself that factoring isn't so bad, but what if the two whole numbers aren't as obvious as the ones we have had so far? Suppose I have no idea what the two numbers are, or even if they exist. I simply start with a factor of 1 and list all the pairs of factors until I find the pair I'm looking for or determine that they do not exist.

The next example illustrates how to find the two numbers.

## EXAMPLE 11

Factor     $8x^2 + 26x + 15$.

Solution:   We need two whole numbers whose product is $(8)(15) = 120$ and whose sum is 26. Assume we have no idea what the numbers are. Start with 1 and list all the pairs of factors of 120 until we find the two whose sum is 26.

The factors of 120 are     1 and 120
2 and 60
3 and 40
4 and 30
5 and 24
6 and 20 and their sum is 26.

Therefore the two numbers are 6 and 20. Since you need the practice, finish the problem from here on your own.

Answer:   $(2x + 5)(4x + 3)$

The following exercise list contains a variety of expressions to be factored. I suggest using the basic strategy as outlined earlier in this unit. Keep in mind that factoring the expressions completely might require repeated factoring.

## EXERCISE

Factor completely:

1.   $7x^2 + 10x + 3$

2.   $2y^2 + 5y - 3$

3.   $6x^2 + 11x + 4$

4.   $3x^3 - 5x^2 - 9x + 15$

5.   $4x^2 - 28x + 48$

6.   $6x^2 + 13x + 6$

7.   $2x^2 + 2x - 24$

8.   $4x^3 - 10x^2 - 6x + 9$

9.   $5x^2 - 4x - 1$

10.   $4x^2 + 8x + 4$

11.   $7x^2 + 13x - 2$

12.   $2x^2 - 7x + 6$

13. $2y^2 - 17y + 35$

14. $7x^2 + 32x - 15$

15. $27x^2z - 3z$

16. $6z^2 + 2z - 4$

17. $x^2z - 16xz + 64z$

18. $6xw^2 + 16wx - 6x$

19. $8y + 4x + 2xy + x^2$

20. $2x^2 + 5x - 2$

21. $8x^2 + 30x - 27$

22. $xy^3 + 2y^2 - xy - 2$

23. $12x^2 - 4x - 5$

24. $x^4 - y^4$

25. $1 - a^4$

--------------------------------------------------------

# UNIT 21

## Solving Quadratic Equations by Factoring

The objective of this unit is to illustrate the approach used to solve equations by factoring. Although the emphasis will be on solving quadratic or second-degree equations, the unit concludes with solving third-degree and higher-degree equations as well.

## QUADRATIC EQUATIONS

> **Definition:** An equation, $ax^2 + bx + c = 0$, with $a$, $b$, and $c$ being real numbers, $a \neq 0$, is called a **second-degree equation** or **quadratic equation**.

In other words, a second-degree or quadratic equation must contain a squared term, $x^2$, and no term with a greater power of $x$.

Examples of second-degree or quadratic equations are:

$$5x^2 + 1 = 0$$
$$x^2 = 25$$
$$x^2 - 5x + 1 = 7$$
$$2x + x^2 = 15x$$

whereas

$x^3 + 2x = 5$ is a third-degree equation

$x^4 = 0$ is a fourth-degree equation

$x^5 + 3x^2 - 5x + 27 = 7x^3$ is a fifth-degree equation

To solve an equation is to find the values of $x$ that satisfy the equation. With a second-degree equation there will be **at most two real solutions**, "at most" meaning there could be

**147**

two unequal real solutions, one solution from two equal real solutions resulting in one value, or no real solution.

There are basically three ways to solve quadratic equations:

1. Factoring—is the fastest method, but not always possible.

2. Completing the square—can be cumbersome and lengthy and so will not be developed in this book.

3. Quadratic formula—will always work, but is long!

---

Definition:   The equation $ax^2 + bx + c = 0$ is called the **standard form** of the quadratic equation.

---

Notice that in the standard form, **all** terms are on the left side of the equal sign, with only 0 on the right side.

The technique I will use to solve a quadratic equation involves four steps. It is based on the principle of zero products, which states that if the product of two numbers is 0, at least one of the numbers must be 0. The four steps are:

1. Write the quadratic equation in standard form. In other words, put **all** terms on the left side, with **only** 0 remaining on the right side of the equal sign.

2. Factor the left side of the equation.

3. Set **each** factor equal to 0.

4. Solve the new equations from step 3.

## EXAMPLE 1

Solve:   $x^2 - 3x = -2$.

- - - - - - - - - - - - - - - - - - - - - - - - - - - - - - - - - - - - -

Solution:                          $x^2 - 3x = -2$

Step 1.      $x^2 - 3x + 2 = 0$

Step 2.    $(x - 1)(x - 2) = 0$

Step 3.      $x - 1 = 0$      or      $x - 2 = 0$

Step 4.          $x = 1$                    $x = 2$

Comment:   Notice that we have **two** solutions. Both must satisfy the original equation. To verify, check by substitution.

$$x^2 - 3x = -2$$

If $x = 1$:      $(1)^2 - 3(1) \stackrel{?}{=} -2$

$$1 - 3 \stackrel{?}{=} -2$$

$$-2 = -2$$

If $x = 2$:      $(2)^2 - 3(2) \stackrel{?}{=} -2$

$$4 - 6 \stackrel{?}{=} -2$$

$$-2 = -2$$

- - - - - - - - - - - - - - - - - - - - - - - - - - - - - - - - - - - - -

## EXAMPLE 2

Solve:   $x^2 = -6 - 5x$.

Solution:                                    $x^2 = -6 - 5x$

      Step 1.     $x^2 + 5x + 6 = 0$

      Step 2.   $(x + 2)(x + 3) = 0$

      Step 3.   $x + 2 = 0$     or     $x + 3 = 0$

      Step 4.     $x = -2$              $x = -3$

Here are two problems for you.

## Problem 1

Solve:   $x^2 - 2x - 48 = 0$.

Solution:

Answer:   $x = 8$ or $x = -6$

## Problem 2

Solve:   $x^2 = 5x - 4$.

Solution:

Answer:   $x = 1$ or $x = 4$

Now let's try several more examples, using the same technique with slight variations.

## EXAMPLE 3

Solve: $-x^2 + 2x + 3 = 0$.

___

Solution: $-x^2 + 2x + 3 = 0$

$\quad\quad\quad x^2 - 2x - 3 = 0$      Note: Multiplying the entire equation by $-1$ makes the factoring easier.

Finish the problem:

Answer: $x = 3$ or $x = -1$

___

## EXAMPLE 4

Solve: $x^2 - 4 = 6x - 13$.

___

Solution: $\quad\quad\quad\quad\quad\quad x^2 - 4 = 6x - 13$

Step 1. $\quad\quad x^2 - 6x + 9 = 0$

Step 2. $\quad (x - 3)(x - 3) = 0$

Step 3. $\quad\quad\quad\quad x - 3 = 0 \quad$ or $\quad x - 3 = 0$

Step 4. $\quad\quad\quad\quad\quad\quad x = 3 \quad\quad\quad\quad\quad x = 3$

There is only one solution since the two factors in step 2 are the same, resulting in two equal roots.

___

## EXAMPLE 5

Solve: $x^2 - 25 = 0$.

___

Solution: $\quad\quad\quad x^2 - 25 = 0$

$\quad\quad\quad (x - 5)(x + 5) = 0$

$\quad\quad\quad x - 5 = 0 \quad$ or $\quad x + 5 = 0$

$\quad\quad\quad\quad x = 5 \quad\quad\quad\quad x = -5$

Alternative solution: $\quad\quad\quad\quad\quad\quad\quad\quad\quad\quad x^2 - 25 = 0$

$\quad\quad\quad\quad\quad\quad\quad\quad\quad\quad\quad\quad\quad\quad\quad\quad x^2 = 25$

Since there is no $x$ term, take the square root of $\quad\quad\quad x = \pm 5$

both sides. Be sure to write both answers,

plus and minus: $\quad\quad\quad\quad\quad\quad\quad\quad\quad\quad x = 5 \quad$ or $\quad x = -5$.

___

## EXAMPLE 6

Solve: $x^2 + 4 = 0$.

___

Solution:   $x^2 + 4 = 0$

$x^2 + 4$ is prime and cannot be factored; hence there is no real solution.

Alternative solution:       $x^2 + 4 = 0$

$$x^2 = -4$$

There is no real solution since there is no real number squared that equals a negative number.

--------------------------------------------------

For some reason many people have difficulty in factoring the next example, where there is no constant term. In reality it is the easiest to factor if you remember to always first factor out any common monomial factor.

====================================================

## EXAMPLE 7

Solve:   $x^2 - 3x = 0$.

--------------------------------------------------

Solution:       $x^2 - 3x = 0$

$$x(x - 3) = 0$$

$x = 0$   or   $x - 3 = 0$

$x = 0$                $x = 3$

--------------------------------------------------

If you have difficulty factoring the next example, don't worry. In Unit 22 I will explain the quadratic formula that allows you to solve a second-degree equation without any factoring.

====================================================

## EXAMPLE 8

Solve:   $5x - 3 = -2x^2$.

--------------------------------------------------

Solution:                   $5x - 3 = -2x^2$

$$2x^2 + 5x - 3 = 0$$

$$2x^2 - x + 6x - 3 = 0$$

$$x(2x - 1) + 3(2x - 1) = 0$$

$$(2x - 1)(x + 3) = 0$$

$2x - 1 = 0$   or   $x + 3 = 0$

$2x = 1$                $x = -3$

$$x = \frac{1}{2}$$

--------------------------------------------------

# HIGHER-DEGREE EQUATIONS

Now we will consider third-degree and higher equations. Recall that with a second-degree equation there are at most two real and different solutions. Similarly, with a third-degree equation there are at most three real solutions, with a fourth-degree equation at most four real solutions, and so on.

The basic technique used to solve these equations is identical to that introduced for quadratic equations and uses the same four steps. The approach allows us to solve equations that are factorable but offers no help for those that are not factorable. A few examples should suffice.

## EXAMPLE 9

Solve: $x^3 - 7x^2 + 6x = 0$.

Solution:

|  | Step 1. | $x^3 - 7x^2 + 6x = 0$ |
|---|---|---|
| Factor out an $x$: | Step 2. | $x(x^2 - 7x + 6) = 0$ |
| Factor again: | Step 2. | $x(x-1)(x-6) = 0$ |
|  | Step 3. | $x = 0, \quad x - 1 = 0, \quad x - 6 = 0$ |
|  | Step 4. | $x = 0 \qquad x = 1 \qquad x = 6$ |

These are the three solutions: $x = 0$, $x = 1$, $x = 6$

## EXAMPLE 10

Solve: $2x^4 - 10x^3 - 28x^2 = 0$.

Solution:  You will save yourself time and energy if you remember to factor out the greatest common monomial factor first.

$$2x^4 - 10x^3 - 28x^2 = 0$$
$$2x^2(x^2 - 5x - 14) = 0$$
$$2x^2(x - 7)(x + 2) = 0$$
$$2x^2 = 0, \quad x - 7 = 0, \quad x + 2 = 0$$
$$x = 0 \qquad x = 7 \qquad x = -2$$

Thus, there are only three solutions to this fourth-degree equation.

Now you try a problem.

## Problem 3

Solve: $x^3 - 3x^2 + 2x = 0$.

Solution:

Answer: $x = 0, 1, 2$

I'll do three more examples.

# EXAMPLE 11

Solve: $x^4 - 16 = 0$.

---

Solution:
$$x^4 - 16 = 0$$
$$(x^2 - 4)(x^2 + 4) = 0$$
$$(x - 2)(x + 2)(x^2 + 4) = 0$$

$$x - 2 = 0, \quad x + 2 = 0, \quad x^2 + 4 = 0$$
$$x = 2 \qquad x = -2 \qquad x^2 = -4$$

no real solution to this part

Thus there are only two real solutions to this fourth-degree equation: $x = 2$, $x = -2$.

---

# EXAMPLE 12

Solve: $2x(2x + 1)(x - 3)(3x - 5) = 0$.

---

Solution: I hope you realize that most of the work has been done for us. The entire left side of the equation has been factored with only a 0 on the right side of the equation; thus we can move immediately to Step 3:

$$2x(2x + 1)(x - 3)(3x - 5) = 0$$

$$2x = 0, \quad 2x + 1 = 0, \quad x - 3 = 0, \quad 3x - 5 = 0$$
$$x = 0 \qquad 2x = -1 \qquad x = 3 \qquad 3x = 5$$
$$x = \frac{-1}{2} \qquad\qquad x = \frac{5}{3}$$

---

# EXAMPLE 13

Solve: $2x^3 - 5x^2 + 7x - 23 = 0$.

---

Solution: This equation cannot readily be factored—grouping does not help, nor is there a common factor. Therefore the best we can do at this time is to say there are at most three real solutions (since it is a third-degree equation), but that we have no idea what they are.

---

To summarize, in Unit 21 we defined, and you should now be able to identify, a second-degree equation as of the type $ax^2 + bx + c = 0$ with $a \neq 0$. Also, you should expect at most two different real solutions. And one approach used to solve these equations is factoring. And to find the solutions you should:

1. Write the equation in standard form.
2. Factor the left side.
3. Set each factor equal to 0.
4. Solve the new equations from step 3.

You should be able to use this same basic technique to solve third-degree and higher-degree equations. This approach allows you to solve equations that are factorable but offers no help for the many equations you are unable to factor.

Before beginning the next unit you should solve the following equations.

## EXERCISES

Solve for $x$:

1. $x^2 + 5x - 14 = 0$

2. $x^2 + 13x + 30 = 0$

3. $x^2 - x + 7 = 0$

4. $4x^2 + 8x + 4 = 0$

5. $x^2 + 5x = 0$

6. $x^2 + 2x = 8$

7. $x^3 + 5x^2 + 6x = 0$

8. $5x^2 - 5x = 0$

9. $2x^2 - 7x + 3 = 0$

10. $2x^2 + 8x + 6 = 0$

11. $z^2 + 4z - 21 = 0$

12. $10x - 10 = 19x - x^2$

13. $3x^2 + 2x = 0$

14. $2 - 2x^2 = 0$

15. $2w^2 + 7w - 4 = 0$

16. $x^3 + 3x^2 - 10x = 0$

17. $(x + 1)(x - 7)(x - 3) = 0$

18. $2x^3 - x^2 + 14x - 7 = 0$

19. $x^4 + 16x^3 + 64x^2 = 0$

20. $12x^2 + 5x - 2 = 0$

# UNIT 22

## Solving Second-Degree Equations—Quadratic Formula

In this unit we will discuss how to use the quadratic formula to solve quadratic equations that cannot be factored.

---

**Quadratic Formula**

Given a quadratic equation $ax^2 + bx + c = 0$, with $a \neq 0$, the solutions are given by

$$x = \frac{-b \pm \sqrt{b^2 - 4ac}}{2a}$$

---

Consider the number under the radical, $b^2 - 4ac$:

1.  If $b^2 - 4ac$ is negative, there are no real solutions since we cannot take the square root of a negative number.

2.  If $b^2 - 4ac = 0$, there are two real and equal solutions.

3.  If $b^2 - 4ac$ is positive, there are two solutions. One is found by using the plus sign, and the other with the minus sign.

Now our technique for solving second-degree equations has been revised to:

1.  Write the equation in standard form.

2.  Factor the left-hand side. (If unable to factor, use the quadratic formula after determining the values for $a$, $b$, and $c$).

Let us assume for the remainder of the unit that you are terrible at factoring and must resort to the quadratic formula at all times!

## EXAMPLE 1

Solve: $3x^2 = x + 2$.

Solution:

$$3x^2 = x + 2$$

Compare with:

$$3x^2 - x - 2 = 0$$
$$ax^2 + bx + c = 0$$

$a$ is the coefficient of the squared term,

$b$ is the coefficient of the first-degree term,

$c$ is the constant,

when the **equation is in standard form.**

Thus $a = 3$, $b = -1$, $c = -2$.

Now, using the formula and substituting:

$$x = \frac{-b \pm \sqrt{b^2 - 4ac}}{2a}$$

$$= \frac{-(-1) \pm \sqrt{1 - 4(-6)}}{6} \quad \text{since } b^2 = (-1)^2 = 1,$$
$$ac = 3(-2) = -6$$

$$= \frac{1 \pm \sqrt{1 + 24}}{6}$$

$$= \frac{1 \pm \sqrt{25}}{6}$$

$$= \frac{1 \pm 5}{6} \quad \text{since } \sqrt{25} = 5$$

Now, to find the two solutions, use first the plus and then the minus:

$$x = \frac{1 + 5}{6} \quad \text{or} \quad x = \frac{1 - 5}{6}$$

$$= \frac{6}{6} \qquad\qquad = \frac{-4}{6}$$

$$= 1 \qquad\qquad = \frac{-2}{3}$$

Note: Had we factored $3x^2 - x - 2 = (x - 1)(3x + 2)$, we would have obtained the same solution.

# EXAMPLE 2

Solve:  $-5x = -3x^2 + 12$.

Solution:
$$-5x = -3x^2 + 12$$

Compare with:
$$\boxed{3}\, x^2 \boxed{-5}\, x \boxed{-12} = 0$$
$$\boxed{a}\, x^2 \boxed{+b}\, x \boxed{+c} = 0$$

Thus $a = 3$, $b = -5$, $c = -12$.

Substitute into the quadratic formula:

$$x = \frac{-b \pm \sqrt{b^2 - 4ac}}{2a}$$

$$= \frac{-(-5) \pm \sqrt{25 - 4(-36)}}{2(3)} \quad \text{since } b^2 = (-5)^2 = 25,$$
$$ac = 3(-12) = -36$$

$$= \frac{5 \pm \sqrt{25 + 144}}{6}$$

$$= \frac{5 \pm \sqrt{169}}{6}$$

$$= \frac{5 \pm 13}{6} \quad \text{since } \sqrt{169} = 13$$

The two solutions are given by:

$$x = \frac{5 + 13}{6} \quad \text{or} \quad x = \frac{5 - 13}{6}$$

$$= \frac{18}{6} \qquad\qquad\quad = \frac{-8}{6}$$

$$= 3 \qquad\qquad\qquad = \frac{-4}{3}$$

Note:   Factoring:

$$3x^2 - 5x - 12 = 0$$

$$(x - 3)(3x + 4) = 0$$

would have led to the same result.

Are you ready to try one?

## Problem 1

Solve: $2x^2 = 5x + 7$.

_ _ _ _ _ _ _ _ _ _ _ _ _ _ _ _ _ _ _ _ _ _ _ _ _ _ _ _ _ _ _ _ _ _ _ _ _

Answer: $x = -1$ or $x = \dfrac{7}{2}$

_ _ _ _ _ _ _ _ _ _ _ _ _ _ _ _ _ _ _ _ _ _ _ _ _ _ _ _ _ _ _ _ _ _ _ _ _

Now study Examples 3–6.

## EXAMPLE 3

Solve: $x^2 - 2x + 4 = 0$.

_ _ _ _ _ _ _ _ _ _ _ _ _ _ _ _ _ _ _ _ _ _ _ _ _ _ _ _ _ _ _ _ _ _ _ _ _

Solution:

$$x^2 \; -2 \; x \; +4 = 0$$

Compare: $a\,x^2 + b \; x \; + c = 0$

Thus $a = 1$, $b = -2$, $c = 4$.

Substitute into the quadratic formula:

$$x = \frac{-b \pm \sqrt{b^2 - 4ac}}{2a}$$

$$= \frac{-(-2) \pm \sqrt{4 - 4(4)}}{2} \quad \text{since } b^2 = (-2)^2 = 4,$$
$$ac = 1(4) = 4$$

$$= \frac{2 \pm \sqrt{4 - 16}}{2}$$

$$= \frac{2 \pm \sqrt{-12}}{2}$$

We can stop right here and write:

No real solution, since we cannot have a negative number under the square root symbol in the real domain.

_____

## EXAMPLE 4

Solve:   $4x^2 - 4x + 1 = 0$.

_____

Solution:          $4\,x^2 - 4\,x + 1 = 0$

Compare:          $a\,x^2 + b\,x + c = 0$

Thus $a = 4$, $b = -4$, $c = 1$.

Substitute into the quadratic formula:

$$x = \frac{-b \pm \sqrt{b^2 - 4ac}}{2a}$$

$$= \frac{-(-4) \pm \sqrt{16 - 4(4)}}{2(4)} \quad \text{since } b^2 = (-4)^2 = 16,$$
$$ac = 4(1) = 4$$

$$= \frac{4 \pm \sqrt{16 - 16}}{8}$$

$$= \frac{4 \pm \sqrt{0}}{8} \quad \text{since the square root of 0 is 0}$$

$$= \frac{4}{8}$$

$$= \frac{1}{2}$$

There are two real and *equal* solutions.

_____

## EXAMPLE 5

Solve: $\frac{1}{3}x^2 - 3x - 2 = 0$.

Solution: Recall from Unit 3 that, whenever we attempted to solve an equation containing fractions, the first step was to simplify. That means to multiply the entire equation by the common denominator of all fractions.

Following that advice with this example, we should first multiply the entire equation by 3 to clear of fractions.

$$3\left(\frac{1}{3}x^2 - 3x - 2\right) = 0$$

$$x^2 - 9x - 6 = 0$$

Compare: $ax^2 + bx + c = 0$

Therefore $a = 1$, $b = -9$, $c = -6$.

Substitute into the quadratic equation:

$$x = \frac{-b \pm \sqrt{b^2 - 4ac}}{2a}$$

$$= \frac{-(-9) \pm \sqrt{81 - 4(-6)}}{2(1)} \quad \text{since } b^2 = (-9)^2 = 81,$$
$$ac = 1(-6) = -6$$

$$= \frac{9 \pm \sqrt{81 + 24}}{2}$$

$$= \frac{9 \pm \sqrt{105}}{2}$$

Or, if you prefer,

$$x = \frac{9 + \sqrt{105}}{2} \quad \text{or} \quad x = \frac{9 - \sqrt{105}}{2}$$

Note: $\left(9 \pm \sqrt{105}\right)/2$ are **exact** answers to the equation. Any attempt to obtain a decimal equivalent for $\sqrt{105}$ by use of a calculator or table will result only in an approximation because $\sqrt{105}$ is irrational.

The application to be made of the answer determines which form should be used.

## EXAMPLE 6

Solve: $3x^2 + x = 6$.

Solution: 
$$3x^2 + x - 6 = 0$$

Compare: $ax^2 + bx + c = 0$

Thus $a = 3$, $b = 1$, $c = -6$.

Substitute:

$$x = \frac{-b \pm \sqrt{b^2 - 4ac}}{2a}$$

$$= \frac{-1 \pm \sqrt{1 - 4(-18)}}{2(3)}$$

$$= \frac{-1 \pm \sqrt{1 + 72}}{6}$$

$$= \frac{-1 \pm \sqrt{73}}{6}$$

Or, if you prefer,

$$x = \frac{-1 + \sqrt{73}}{6} \qquad \text{or} \qquad x = \frac{-1 - \sqrt{73}}{6}$$

It's your turn again.

## Problem 2

Solve:  $3x^2 + x - 1 = 0$.

Answer:  $x = \dfrac{-1 \pm \sqrt{13}}{6}$

Here are two more examples to end the unit.

## EXAMPLE 7

Solve:   $x^2 - 4x - 7 = 0$.

Solution:   Skipping a few steps and going right to the quadratic formula with $a = 1$, $b = -4$, $c = -7$:

$$x = \frac{-b \pm \sqrt{b^2 - 4ac}}{2a}$$

$$= \frac{-(-4) \pm \sqrt{16 - 4(-7)}}{2(1)}$$

$$= \frac{4 \pm \sqrt{16 + 28}}{2}$$

$$= \frac{4 \pm \sqrt{44}}{2}$$

At this point you have three choices:
1.   Stop and leave the answer as above.
2.   Use a calculator to convert to decimal approximations.
3.   Simplify the answer further as follows:

$$x = \frac{4 \pm \sqrt{4 \cdot 11}}{2}$$

$$= \frac{4 \pm 2\sqrt{11}}{2} \quad \text{since } \sqrt{4} = 2$$

$$= \frac{2(2 \pm \sqrt{11})}{2} \quad \text{by factoring out 2}$$

$$= \frac{\cancel{2}(2 \pm \sqrt{11})}{\cancel{2}} \quad \text{and cancelling}$$

$$= 2 \pm \sqrt{11} \quad \text{is the answer in simplest form.}$$

## EXAMPLE 8

Solve:   $x^3 + x^2 - x = 0$.

Solution:   $x^3 + x^2 - x = 0$

$x(x^2 + x - 1) = 0$

Since $x^2 + x - 1$ cannot be readily factored, go to step 3.

$x = 0, \quad x^2 + x - 1 = 0$

This is now a second-degree equation, and the quadratic formula can be used.

Since $x^2 + x - 1 = 0$
and $ax^2 + bx + c = 0$, then $a = 1$, $b = 1$, $c = -1$.

Substituting into the quadratic formula gives

$$x = \frac{-b \pm \sqrt{b^2 - 4ac}}{2a}$$

$$= \frac{-1 \pm \sqrt{1 - 4(-1)}}{2}$$

$$= \frac{-1 \pm \sqrt{5}}{2}$$

There are three solutions:

$$x = 0, \qquad x = \frac{-1 + \sqrt{5}}{2}, \qquad x = \frac{-1 - \sqrt{5}}{2}$$

---

When attempting to solve third-degree or higher-degree equations, if we are unable to factor the expression, there is no formula to help us. Unlike the quadratic formula used to solve second-degree equations, no cubic formula exists for solving third-degree or higher-degree equations.

You now should be able to solve any second-degree or quadratic equation. The procedure is to write the equation to be solved in standard form. Then, if the left side cannot be readily factored, compare it with $ax^2 + bx + c = 0$ to determine the values of $a$, $b$, and $c$. These values are then substituted into the quadratic formula

$$x = \frac{-b \pm \sqrt{b^2 - 4ac}}{2a}$$

Now try to solve the following equations using the quadratic formula.

## EXERCISES

Solve for $x$:

1. $x^2 + 3x - 1 = 0$

2. $2x^2 - 3x - 2 = 0$

3. $4x^2 + 4x - 4 = 0$

4. $2x^2 - 3x - 1 = 0$

5. $x^2 + x - 20 = 0$

6. $\frac{1}{5}x^2 - 5x + 1 = 0$

7.  $4 = x^2 - 3x$

8.  $x^2 - 2x + 2 = 0$

9.  $2x^2 - x = 0$

10.  $4x^2 - 4x + 1 = 0$

11.  $2x^2 + 7x + 9 = 0$

12.  $x^2 - 2x - 10 = 0$

13.  $3x^2 + x - 3 = 0$

14.  $2x^2 + \sqrt{3}x - 4 = 0$

15.  $x^5 - x^4 - 12x^3 = 0$

16.  $2x^4 + 2x^3 + 2x^2 = 0$

# UNIT 23

# Graphing Linear Equations in Two Variables

The purpose of this unit is to provide you with an understanding of how to graph linear equations. When you have finished the unit, you will be able to identify linear equations in two variables, distinguish them from other types of equations, and graph them.

## RECTANGULAR OR CARTESIAN COORDINATE SYSTEM OF GRAPHING

We will use a rectangular or Cartesian coordinate system for all graphs. You probably remember the basic facts:

Usually the horizontal axis is labeled $x$.
The vertical axis is labeled $y$.
The origin, $O$, is the point where the axes cross.
Each of the four sections is called a quadrant.
Quadrants are numbered I, II, III, and IV as shown below.

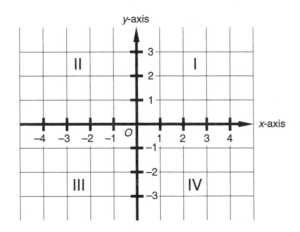

The coordinates of a point are written as an ordered pair, $(x, y)$, where $x$ is the number of horizontal units the point is from the origin:

> to the right if $x$ is positive,
> to the left if $x$ is negative;

and $y$ is the number of vertical units the point is from the origin:

> up if $y$ is positive,
> down if $y$ is negative.

## EXAMPLE 1

Plot:  $A(2, 3)$.

Solution:  1.  Start at the origin.

2.  The first number, 2, is positive. Move two spaces to the right.

3.  Next, because the second number, 3, is positive, move three spaces up.

4.  Draw a dot and label the point.

Answer:

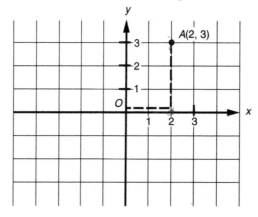

## EXAMPLE 2

Plot:  $B(-3, 1)$.

Solution:  1.  Start at the origin.

2.  The first number, $-3$, is negative. Move three spaces to the left.

3.  Next, because the second number, 1, is positive, move one space up.

4.  Draw a dot and label the point.

Answer:

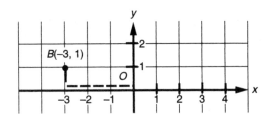

---

## EXAMPLE 3

Plot:   C(2.5, –2).

---

Solution:   1.   Start at the origin.

2.   Move 2.5 spaces to the right because $x$ is 2.5.

3.   Then move two spaces down because $y$ is –2.

4.   Draw a dot and label the point.

Answer:

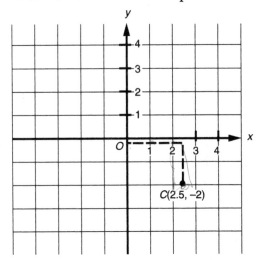

---

When plotting points yourself, there is no need to draw the dotted lines as I did in the examples. That was done only to show the procedure.

Now try to plot some points yourself. Use the blank grid provided. Be sure to label each point.

Problem 1     $D(-4, 3)$.

Problem 2     $E(5, 2)$.

Problem 3     $F\left(\dfrac{3}{2}, -3\right)$.

Problem 4     $G(-2, 0)$.

Problem 5     $H(0, -5)$.

Problem 6     $J(-4.5, -2)$.

Answer:

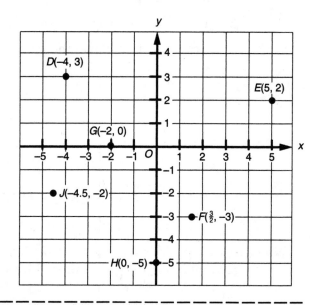

Note:  Each ordered pair corresponds to one point on the graph, and for each point on the graph there is one ordered pair.

Now do a problem in which the question is reversed.

## Problem 7

Find the coordinates for each of the points on the graph below:

A(  ,  )
B(  ,  )
C(  ,  )
D(  ,  )
E(  ,  )

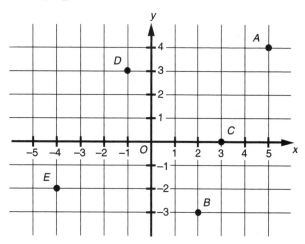

Answers:   *A*(5, 4)   *B*(2, –3)   *C*(3, 0)   *D*(–1, 3)   *E*(–4, –2)

Given any ordered pair, $(x, y)$, you should now be able to plot it. Remember that the procedure is **to start at the origin and move according to the following guide:**

**First:** if $x$ is positive, move to the right;
if $x$ is negative, move to the left.

**Next:** if $y$ is positive, move up;
if $y$ is negative, move down.

**Then draw a dot and label the point.**

Now we will consider what is meant by a linear equation in two variables, how we can identify such equations, and what procedure we can use to graph them.

---

Definition:   An equation that has the form $ax + by = c$, with $a$, $b$, and $c$ being real numbers, $a$ and $b$ not both zero, is **a linear equation in two variables**.

---

In other words, in a linear equation in two variables:

1. There are one or two variables.

2. Each variable is involved in only the four fundamental operations—addition, subtraction, multiplication, and division.

3. Neither variable is raised to any power other than 1.

4. Neither variable appears in any denominator.

5. No term contains a product of the two variables.

Here are some examples of linear equations in two variables:

$$3x + 2y = 5$$

$$\frac{1}{2}y + 7x = -\sqrt{13}$$

$$x - y = 0$$

$$\frac{x}{3} = 11y - 201$$

Here are some examples that are *not* linear equations in two variables:

$$3x^2 + 2y = 7$$

$$-5x + \sqrt{y} = 10$$

$$\frac{1}{x} - 7 = y$$

$$-13xy + y = 321$$

$$-2x - 3y + z = 15$$

Before proceeding, determine why each of the above is *not* a linear equation in two variables.

In the rest of this unit, we will consider only linear equations in two variables.

# GRAPHING LINES

Now we are ready to begin graphing linear equations.

---

Definition:    The **graph of an equation** is the set of all points whose coordinates satisfy the equation.

---

The graph can be thought of as a "picture" of the solution set of the equation. The graph of a linear equation is a straight line.

To graph a linear equation, locate three points whose coordinates satisfy the equation and connect them with a straight line. Actually only two points are needed; the third point serves as a check.

### Plotting Points

One basic procedure for graphing a linear equation in two variables is by plotting points. To locate each of the three points:

1.  Select some convenient value for $x$.
2.  Substitute this value into the equation.
3.  Solve for $y$.

---

## EXAMPLE 4

Graph:   $x + 2y = 6$.

Solution: $x + 2y = 6$ is a linear equation; therefore its graph will be a straight line.

Locate the first point:

1. Select a convenient value for $x$.     Let $x = 2$.
2. Substitute into the equation.        $2 + 2y = 6$
3. Solve for $y$.                               $2y = 4$
                                                       $y = 2$

Then (2, 2) is a point on the graph of the equation because it satisfies the equation:

$$2 + 2(2) \overset{?}{=} 6$$
$$6 = 6$$

Repeat the procedure to find a second point:

1. Select a value for $x$.     Let $x = 0$.
2. Substitute.                      $0 + 2y = 6$
3. Solve for $y$.                        $y = 3$

Thus (0, 3) is a second point on the graph because (0, 3) satisfies the equation:

$$0 + 2(3) \overset{?}{=} 6$$
$$6 = 6$$

Repeat the procedure to find a third point:

1. Select a value for $x$.     Let $x = 4$.
2. Substitute.                      $4 + 2y = 6$
3. Solve for $y$.                      $2y = 2$
                                              $y = 1$

Then (4, 1) is a third point on the graph because (4, 1) satisfies the equation:

$$4 + 2(1) \overset{?}{=} 6$$
$$6 = 6$$

Plot the three points and connect with a straight line. Be sure to draw arrows at each end of the line to indicate that it extends indefinitely in both directions.

Answer:

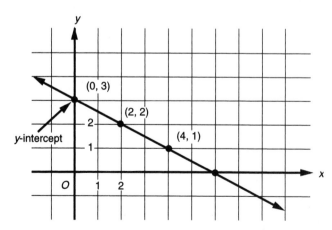

Note that there was nothing special about the three values selected for $x$. It often happens that $x = 0$ is a convenient value to use. The other two were selected so that when I divided by 2 in the final step, the result would be a whole number.

---

The point where the line crosses the $y$-axis is called the $y$-intercept. In other words, the $y$-intercept is the value of $y$ when $x = 0$. In the above example the $y$-intercept is 3.

## EXAMPLE 5

Graph: $3x - y = 5$.

Solution: $3x - y = 5$ is a linear equation; the graph will be a straight line. Locate three points on the line.

Let $x = 0$; then $3(0) - y = 5$,
$$0 - y = 5,$$
$$-y = 5,$$
$$y = -5 \text{ and } (0, -5) \text{ is one point.}$$
Let $x = 2$; then $3(2) - y = 5$,
$$6 - y = 5,$$
$$1 = y \text{ and } (2, 1) \text{ is a second point.}$$
Let $x = -2$; then $3(-2) - y = 5$,
$$-6 - y = 5,$$
$$-y = 11,$$
$$y = -11 \text{ and } (-2, -11) \text{ is a third point.}$$

Plot the three points, and connect them with a straight line.

Answer:

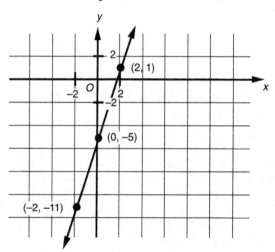

Here's a problem for you.

## Problem 8

Question: What is the $y$-intercept of Example 5?

Solution:

------------------------------------------------

### Creating a Table of Values

Have you noticed that in each of the last two examples we ended up solving the same equation three times? Wouldn't it be faster if it was necessary to solve it only once? This is the basis of our next few examples. This time, **before** selecting convenient values for $x$, solve the equation for $y$ and create a table of values by substitution. As a refresher on how to solve an equation for $y$, refer to Unit 6 on solving literal equations.

To graph a linear equation using a table of values:

1.  First solve the equation for $y$.

2.  Select some convenient values for $x$. Then, by substitution, create a table of values whose coordinates satisfy the equation.

3.  Plot the points on the graph and connect them with a straight line.

## EXAMPLE 6

Graph:   $2x + y = 3$.

------------------------------------------------

Solution:   $2x + y = 3$ is a linear equation; the graph will be a straight line.

1.  Solve the equation for $y$.     $2x + y = 3$

$$y = -2x + 3$$

2.  Create a table of values.

Select three convenient values for $x$ and solve for $y$ by substitution.

Let $x = -2$; then $y = -2(-2) + 3 = 4 + 3 = 7$
Let $x = 0$; then $y = -2(0) + 3 = 0 + 3 = 3$
Let $x = 1$; then $y = -2(1) + 3 = -2 + 3 = 1$

Our table of values would look like the following, with the corresponding points to the right.

| $x$ | $y$ | $(x, y)$ |
|---|---|---|
| –2 | 7 | (–2, 7) |
| 0 | 3 | (0, 3) |
| 1 | 1 | (1, 1) |

3.  Plot the three points and connect them with a straight line.

Answer:

- - - - - - - - - - - - - - - - - - - - - - - - - - - - - - -

I'll do another example.

## EXAMPLE 7

Graph:   $2x + 3y = 21$.

- - - - - - - - - - - - - - - - - - - - - - - - - - - - - - -

Solution:   $2x + 3y = 21$ is a linear equation; the graph will be a straight line.

1.  Solve the equation for $y$.

$$2x + 3y = 21$$
$$3y = -2x + 21$$
$$y = \frac{-2x + 21}{3}$$
$$= \frac{-2}{3}x + 7$$

2.  Create a table of values.
    Select three convenient values for $x$ and solve for $y$ by substitution.

$$\text{Let } x = 0; \text{ then } y = \frac{-2(0)}{3} + 7 = 0 + 7 = 7$$

$$\text{Let } x = 3; \text{ then } y = \frac{-2(3)}{3} + 7 = -2 + 7 = 5$$

$$\text{Let } x = -2; \text{ then } y = \frac{-2(-2)}{3} + 7 = \frac{4}{3} + 7 = 1\frac{1}{3} + 7 = 8\frac{1}{3}$$

Our table of values looks like the following, with the corresponding points to the right.

| $x$ | $y$ | $(x, y)$ |
|-----|-----|----------|
| 0 | 7 | $(0, 7)$ |
| 3 | 5 | $(3, 5)$ |
| $-2$ | $8\frac{1}{3}$ | $\left(-2, 8\frac{1}{3}\right)$ |

3. Plot the three points, and connect them with a straight line.

Answer:

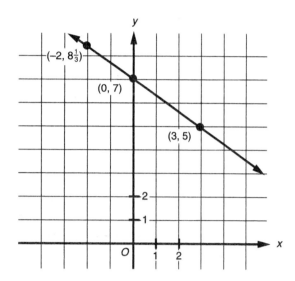

---

Now you try to graph a linear equation.

---

## Problem 9

Graph: $5x + 2y = 10$.

---

Solution: $5x + 2y = 10$ is a linear equation; the graph will be a straight line.

1. Solve the equation for $y$.

2. Create a table of values:

| $x$ | $y$ | $(x, y)$ |
|---|---|---|
| 0 | | ( , ) |
| 2 | | ( , ) |
| −2 | | ( , ) |

Have you noticed that by first solving the equation for $y$, it is easier to decide what values of $x$ are most convenient; that is, you know in advance if you want to select numbers divisible by 3 as in Example 7 or divisible by 2 as in Problem 9. Also, once you have selected the values for $x$, you need only to use substitution to find the corresponding $y$-values.

3. Plot the three points on the blank grid provided and connect them with a straight line.

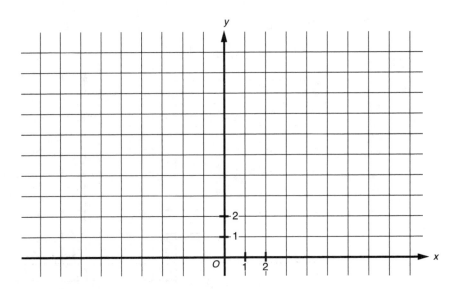

Reminder:   Be sure to put the arrows at each end to indicate that the line continues on in both directions.

Answer:

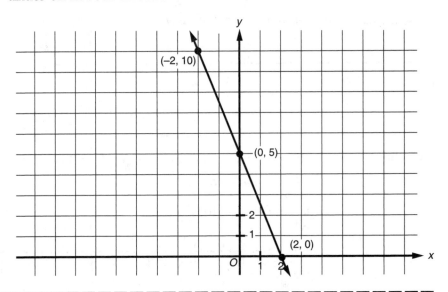

Try the next graph without any clues.

## Problem 10

Graph:   $-x + y = 7$. Identify the $y$-intercept.

Solution:

Answer:

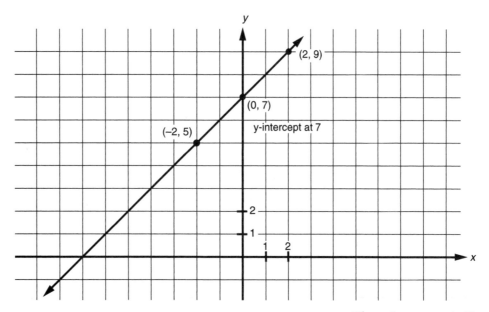

The *y*-intercept is 7.

The point where the line crosses the *x*-axis is called the *x*-intercept. In other words, the *x*-intercept is the value of *x* when $y = 0$.

## EXAMPLE 8

Find the *x*- and *y*-intercepts for $5x + 2y = 10$.

Solution:   The *x*-intercept is the value of *x* when $y = 0$.

By substitution:   $5x + 2(0) = 10$

$$5x = 10$$

$$x = 2$$

The *x*-intercept is 2.

Recall that the $y$-intercept is the value of $y$ when $x = 0$.

By substitution: $5(0) + 2y = 10$

$$2y = 10$$

$$y = 5$$

The $y$-intercept is 5.

Refer to the graph of Problem 9 to verify that the line crosses the $x$-axis at 2 and the $y$-axis at 5.

---

Often the $x$- and $y$-intercepts are wise choices for two of the three points used to graph a line.

Whenever possible, a linear equation should be simplified *before* locating the three points used to graph it.

## EXAMPLE 9

Graph: $x + 3y = 9$.

---

Solution:

It is a linear equation; the graph will be a straight line. Locate three points.

I will start with the intercepts.

The $y$-intercept is where $x = 0$, $\quad 0 + 3y = 9$

$$y = 3$$

Therefore the $y$-intercept is 3.

The $x$-intercept is where $y = 0$, $\quad x + 3(0) = 9$

$$x = 9$$

Therefore the $x$-intercept is 9. Locate one additional point as a check.

Let $x = 3$; then $3 + 3y = 9$,

$$3y = 6,$$

$$y = 2 \text{ and the point is } (3, 2).$$

Answer:

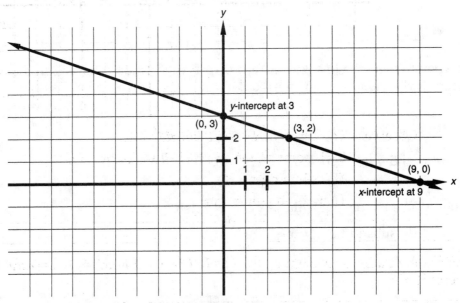

# TWO SPECIAL TYPES OF LINEAR EQUATIONS

There are two special types of linear equations in which only a single variable, either $x$ or $y$, is involved. The graph of such an equation will be either a vertical or a horizontal line, depending on which variable is present.

## EXAMPLE 10

Graph: $x = 3$.

Solution: Recall that a linear equation is of the form $ax + by = c$ with $a$ and $b$ not both zero.

Compare:
$$\boxed{a}x \boxed{+ b}y = \boxed{c}$$
$$\boxed{\;}x \boxed{\;} = \boxed{3} \quad \text{so } a = 1, b = 0, c = 3.$$

Therefore $x = 3$ is classified as a linear equation; its graph will be a straight line. Locate three points whose coordinates satisfy the equation.

For the moment, think of the equation as $x + 0y = 3$. Three points whose coordinates satisfy the equation are $(3, 2)$, $(3, 5)$, and $(3, 0)$.

To verify $(3, 2)$:
$$x + 0y = 3$$
$$3 + 0(2) = 3$$
$$3 = 3$$

Thus point $(3, 2)$ is on the line. Verify the other two points yourself.

Answer:

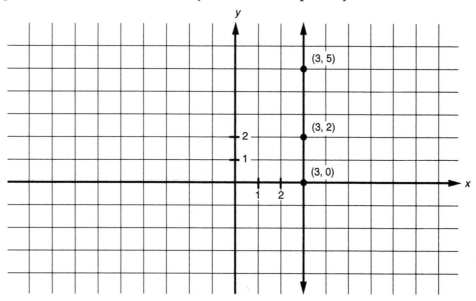

Observe: The graph of $x = 3$ is a vertical line.

Each point on the line has an $x$-coordinate of 3.

The $x$-intercept is 3.

There is no $y$-intercept.

---

The graph of a linear equation $x = r$, where $r$ is a real number, is a *vertical line* with $x$-intercept of $r$.

## EXAMPLE 11

Graph: $x = -4$.

Solution: The graph of a linear equation $x = -4$ is a vertical line with $x$-intercept of $-4$.

Answer:

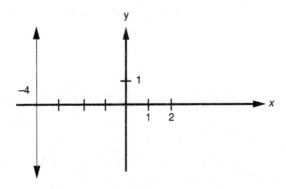

## EXAMPLE 12

Graph: $y = 2$.

Solution: By a similar line of reasoning,
compare: $\boxed{a} x \boxed{+ b} y = \boxed{c}$
$y = \boxed{2}$ so $a = 0$, $b = 1$, $c = 2$.

Thus $y = 2$ is also a linear equation; its graph will be a straight line.

If you think of the equation as $0x + y = 2$, three points whose coordinates satisfy the equation are $(5, 2)$, $(-3, 2)$, and $(0, 2)$. This time the graph is a horizontal line with each point having a $y$-coordinate of 2. The $y$-intercept is 2.

Answer:

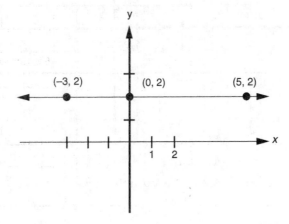

---

The graph of a linear equation $y = p$, where $p$ is a real number, is a *horizontal line* with $y$-intercept of $p$.

You do a final problem.

## Problem 11

Graph: $\left.\begin{array}{l} x = 5 \\ y = 4 \\ y = -3 \end{array}\right\}$ on the same set of axes.

Solution:

Answer:

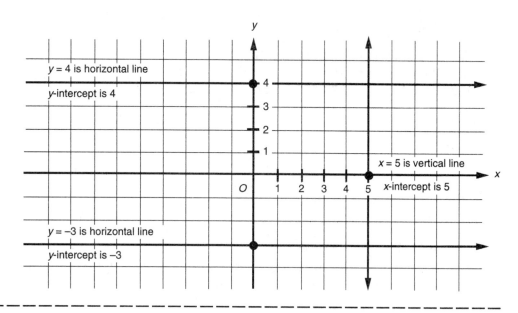

# ALGEBRA AND THE CALCULATOR (Optional)

### Graphing Lines

So that we start at the same place, after pressing the ON key,

press CLEAR MODE CLEAR and

press 2nd TblSet to view the table setup menu.

Now assign  TblStart = 0  by pressing 0 ENTER.

$\Delta$ Tbl = 1  by pressing 1 ENTER.

Indpnt: Ask  by highlighting Ask and pressing ENTER.

Depend: Auto  by highlighting Auto and pressing ENTER.

You should see the following:

```
TABLE SETUP
 TblStart=■
 ₔTbl=1
Indpnt: Auto Ask
Depend: Auto Ask
```

To return to the home screen, press 2nd QUIT or CLEAR.

For now we will use the standard viewing window for our graphs. Its dimensions are [−10, 10] by [−10, 10]. That is, the x-axis will be shown from −10 to 10 and the y-axis will be shown from −10 to 10 as pictured at the right.

To use the standard viewing window, press Y= CLEAR ZOOM 6, and a blank standard viewing window should appear on the screen.

Press WINDOW to confirm that the viewing window's dimensions agree with the screen shown here. If they don't, reset to match.

Press 2nd FORMAT and select RECTGC to use rectangular coordinates.

Return to the home screen by pressing CLEAR or 2nd QUIT.

```
WINDOW
 Xmin=-10
 Xmax=10
 Xscl=1
 Ymin=-10
 Ymax=10
 Yscl=1
 Xres=1
```

### Creating a Table of Values and the Graph of a Line

## EXAMPLE 13

Graph:   $y = 1.3x + 2.75$ using a graphing calculator.

Solution:   This example is the same as what you have been doing except that the equation includes decimals. Thus, manual calculations would be tedious and time-consuming, and the graph would be an approximation. We will do it on a calculator.

To enter the equation, press the following keys.

Y= CLEAR 1 · 3 X,T,θ,n + 2 · 7 5 ENTER.

You should see the following.

To find points on the line, press 2nd TABLE.

Select a value for $x$.   Let $x = 0$.

To find $y$, press 0 ENTER.

Repeat by entering 1, 2, and 3.

Reminder: The $\boxed{(-)}$ key is used to denote a negative number. An error message will appear if the subtraction key $\boxed{-}$ is used by mistake. Finish the above example by entering –1, –2, and –3.

You should see the following.

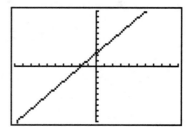

To see the graph, press $\boxed{\text{GRAPH}}$. Surprisingly easy, wasn't it?

---

## EXAMPLE 14

Graph:   $y = -\dfrac{1}{2}x + 4$ using a graphing calculator.

Find:   a.  $y$ when $x = 3.74$ using the $\boxed{\text{CALC}}$ key.

   b.  the $y$-intercept using the $\boxed{\text{TRACE}}$ key.

   c.  the $x$-intercept using the $\boxed{\text{CALC}}$ key.

---

Solution:   Create a table of values for the equation using the following keystrokes:

$\boxed{\text{Y=}}$ $\boxed{\text{CLEAR}}$ $\boxed{(}$ $\boxed{(-)}$ $\boxed{1}$ $\boxed{\div}$ $\boxed{2}$ $\boxed{)}$ $\boxed{\text{X,T,}\theta\text{,n}}$ $\boxed{+}$ $\boxed{4}$ $\boxed{\text{ENTER}}$.

To see various $x$- and $y$-values, press $\boxed{\text{2nd}}$ $\boxed{\text{TABLE}}$.

To see the graph, press GRAPH.

a. With the graph on the screen, to find $y$ when $x = 3.74$ using the CALC key, press 2nd CALC 1 3 · 7 4 ENTER. The equation appears at the upper right, the $x$- and $y$-values of 3.74 and 2.13 are at the bottom, and a star appears on the graph at the point with coordinates (3.74, 2.13).

b. To find the $y$-intercept using the TRACE key, press TRACE. Using only the ▶ and ◀ arrows, move the blinking star (cursor) along the graph until the $x$-value is 0. The $y$-intercept, in this example, $y = 4$, appears at the bottom of the screen.

c. To find the $x$-intercept, also called a zero, using the CALC key requires a bit more work. Press 2nd CALC 2.
At the <u>Left Bound?</u> prompt, using only the ▶ and ◀ arrows, move the cursor to a point just left of the $x$-intercept and press ENTER. If you find it difficult to decide where to place the cursor, keep in mind that both the $x$- and $y$-values will be positive.
At the <u>Right Bound?</u> prompt, move the cursor to a point just to the right of the $x$-intercept. Here $x$ will be positive and $y$ will be negative; then press ENTER.

At the <u>Guess?</u> prompt, move the cursor between the boundaries shown on the screen by the two arrows, ▶ ◀, and press ENTER. The answer appears at the bottom of the screen: zero $x = 8$ $y = 0$.

------------------------------------------------------------

Before continuing, I suggest you redo Examples 4–9 for practice with your calculator. Also, when entering functions, experiment with the DEL and INS keys.

============================================================

## EXAMPLE 15

Graph $2x + 5y = 60$ using a graphing calculator.

------------------------------------------------------------

Solution: Before the equation can be entered into the calculator, it must be solved for $y$ as follows:

$$2x + 5y = 60$$
$$5y = -2x + 60$$
$$y = -\frac{2}{5}x + 12$$

Press Y= ( (−) 2 ÷ 5 ) X,T,θ,n + 1 2 ENTER. This time, however, when we press GRAPH, only a small portion of the line is visible in the standard viewing window. Obviously the viewing window needs to be enlarged. Several methods are available.

One method is to press TRACE ZOOM 3 ENTER.

If you are curious about the new dimensions, press WINDOW.

A second method is to manually reset the dimensions of the window. For example, you might find the $x$- and $y$-intercepts, which are 30 and 12, respectively, and reset the window's dimensions accordingly.

To reset the dimensions of the window, press WINDOW and reset Xmax to 35 to accommodate the $x$-intercept of 30, and reset Ymax to 15 to accommodate the $y$-intercept of 12. Press GRAPH to view the graph in the new enlarged viewing window.

Now I encourage you to do a bit of experimenting with the settings and the various keys. For instance, press 2nd FORMAT, select various options, and observe the results on a graph. Or, change the viewing window's dimensions again. ZOOM 6 will always take you back to the standard viewing window. Or, use TRACE to move along the line and observe what happens when you reach the edge of the viewing window.

You should now be able to recognize and graph linear equations in two variables by one of three methods: plotting points, creating a table of values, or graphing the $x$- and $y$-intercepts.

To plot points:

select three convenient values for $x$;
substitute these values into the equation and solve for $y$;
plot the three points and connect them with a straight line.

To create a table of values:

solve the equation for $y$;
select three convenient values for $x$ and solve for $y$ by substitution.

To graph using the $x$- and $y$-intercepts:

locate the $y$-intercept, which is the value of $y$ when $x = 0$;
locate the $x$-intercept, which is the value of $x$ when $y = 0$;
plot and connect the two points with a straight line.

Before beginning the next unit you should graph the following equations.

## EXERCISES

Identify the $x$-intercept, the $y$-intercept, and graph each equation:

1.  $x + y = 8$

2.  $3x - 4y = 12$

3.  $7x + y = 10$

4.  $y = 2$

5. $y - 2x = 0$

6. $x + 3y = 21$

7. $y = \dfrac{2}{7}x + 4$

8. $x = -3$

9. $-3x = 2y + 4$

10. $y = \dfrac{2}{5}x - 1$

C11. Graph $y = -3.14x + 7.92$ using a graphing calculator.

C12. Using a graphing calculator and given $y = 1.23x + 5.76$

    a.   Graph the linear equation.

    b.   Find the $y$-intercept.

    c.   Find the $x$-intercept.

    d.   Find $y$ so that the point $(-1.12, y)$ is on the line.

# UNIT 24

## Graphing Quadratic Equations

In this unit you will learn to recognize quadratic equations in two variables and to graph them using a minimum number of well-chosen points.

> **Definition:** $y = ax^2 + bx + c$, with $a$, $b$, and $c$ being real numbers, $a \neq 0$, is called a **quadratic** or **second-degree equation in two variables.**

In other words, a second-degree equation in two variables must contain two variables, a squared term, $x^2$, and no higher powered term. In this unit we will consider only second-degree or quadratic equations.

There are basically two ways to graph quadratic equations:

1. Find and plot a large number of points—this can be time-consuming, and key points are often missed.

2. Plot a few well-chosen points based on knowledge of quadratic equations—this is the approach we will use.

Before continuing, let me illustrate the first approach, using the simplest possible second-degree equation in two variables. Recall from Unit 23 that the graph of an equation was defined as the set of all points $(x, y)$ whose coordinates satisfy the equation.

## EXAMPLE 1

Graph: $y = x^2$.

Solution:  Find and plot a large number of points whose coordinates satisfy the equation.

Let $x = -2$; then $y = (-2)^2 = 4$, which gives point $(-2, 4)$.

Let $x = -1$; then $y = (-1)^2 = 1$, which gives point $(-1, 1)$.

Let $x = -\dfrac{1}{2}$; then $y = \left(-\dfrac{1}{2}\right)^2 = \dfrac{1}{4}$, which gives $\left(-\dfrac{1}{2}, \dfrac{1}{4}\right)$.

Let $x = 0$; then $y = (0)^2 = 0$, which gives $(0, 0)$.

Let $x = \dfrac{1}{2}$; then $y = \left(\dfrac{1}{2}\right)^2 = \dfrac{1}{4}$, which gives $\left(\dfrac{1}{2}, \dfrac{1}{4}\right)$.

Let $x = 1$; then $y = 1^2 = 1$, which gives $(1, 1)$.

Let $x = 2$; then $y = 2^2 = 4$, which gives $(2, 4)$.

Answer:

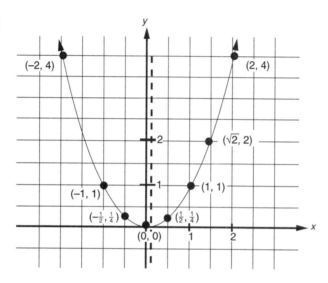

If you are not convinced that this is the graph, locate as many more points as you need to be sure. You might let $x$ equal such values as 1.5, 3, –4, or even $\sqrt{2}$, and plot the corresponding points.

---

Be sure to notice that every ordered pair whose coordinates satisfy the equation, $y = x^2$, must correspond to some point on the curve. And every point on the curve must have coordinates that satisfy the equation.

As you probably observed, plotting points could become quite time-consuming if the quadratic equation were more complicated, so we will move to the second method.

Definition:  $y = ax^2 + bx + c$ is called the **standard form** of a quadratic equation.
Note:  **All** terms are on the right side of the equal sign with **only** $y$ on the left.

Fortunately, the graph of a quadratic equation is very predictable. The following are some basic facts that can be used when graphing a quadratic equation in two variables.

# BASIC FACTS ABOUT THE GRAPH OF A QUADRATIC EQUATION IN STANDARD FORM: $y = ax^2 + bx + c$

1. The graph of a quadratic equation is a smooth, $\cup$-shaped curve called a parabola.
2. If $a$, the coefficient of the squared term, is positive, the curve opens up: $\cup$.
   If $a$ is negative, the curve opens down: $\cap$.
3. The $y$-intercept is $c$, the constant. Remember that the $y$-intercept is the value of $y$ when $x = 0$; thus $y = a(0)^2 + b(0) + c = c$.
4. There are **at most two x-intercepts**, "at most" meaning there can be two, one, or none. The $x$-intercepts are the values of $x$ when $y = 0$; thus the solution to $0 = ax^2 + bx + c$ yields the $x$-intercepts.
5. The low point on the curve (or the high point if the curve opens down) is called the **vertex**.
6. The vertex is located at the point

$$\left( \frac{-b}{2a}, \frac{4ac - b^2}{4a} \right).$$

7. The curve is symmetric to a vertical line through the vertex. The vertical line is called the **axis of symmetry**. Its equation is $x = \dfrac{-b}{2a}$.

On the basis of the above information about quadratic equations in two variables, it is now possible to graph such equations using only a few well-chosen points. By "few" I mean plotting at least three points, but no more than seven points.

# SUGGESTED APPROACH FOR GRAPHING A QUADRATIC EQUATION IN TWO VARIABLES

1. Write the quadratic equation in standard form.
2. Determine whether the parabola opens up or down.
3. Find the $y$-intercept at $c$.
4. Solve $0 = ax^2 + bx + c$ to find the $x$-intercepts, if the equation is easily factored.
5. Find the coordinates of the vertex:

$$\left( \frac{-b}{2a}, \frac{4ac - b^2}{4a} \right).$$

6.  Draw the axis of symmetry with a dashed line. It is a vertical line through the vertex.
7.  Locate one point on either side of the vertex if necessary.
8.  Plot the above points, and connect them with a smooth, $\smile$-shaped curve.

   Examples 2–4 illustrate this approach.

## EXAMPLE 2

Graph:   $y = x^2 - 5x + 4$.

Solution:   1.  Write in standard form and compare:
$$y = x^2 - 5x + 4$$
$$y = ax^2 + bx + c$$
Thus $a = 1$, $b = -5$, $c = 4$.

Graph will be a parabola.

2.  Since $a$ is positive, the curve opens up: $\smile$.
3.  The $y$-intercept is $c = 4$.
4.  Find the $x$-intercepts by solving:
$$0 = x^2 - 5x + 4$$
$$0 = (x - 4)(x - 1)$$
$$x = 4, \qquad x = 1$$

The $x$-intercepts are 4 and 1.

5.  Find the coordinates of the vertex:
$$x = \frac{-b}{2a} = \frac{-(-5)}{2(1)} = \frac{5}{2}$$
$$y = \frac{4ac - b^2}{4a} = \frac{4(1)(4) - (-5)^2}{4(1)} = \frac{16 - 25}{4} = \frac{-9}{4}$$

The vertex is located at $\left(\dfrac{5}{2}, \dfrac{-9}{4}\right)$.

6.  Draw the axis of symmetry with a dashed line. It is a vertical line through the vertex, $\left(\dfrac{5}{2}, \dfrac{-9}{4}\right)$.

7.  Locate one point on either side of the vertex. Since the vertex is at $x = \dfrac{5}{2} = 2.5$:
$$\text{let } x = 2; \text{ then } y = (2)^2 - 5(2) + 4 = 4 - 10 + 4 = -2$$
$$\text{let } x = 3; \text{ then } y = (3)^2 - 5(3) + 4 = 9 - 15 + 4 = -2$$

8.  Plot the above points and connect them with a smooth curve.

Answer:

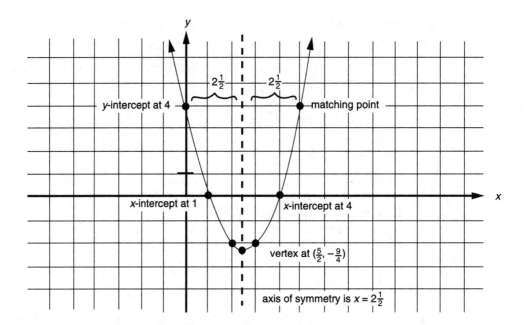

Before going on to another example, Let's mention a few points about the above graph. The axis of symmetry is indicated with a dashed line. Because the curve is symmetrical about this line, there must be a point $2\frac{1}{2}$ spaces away that corresponds to the $y$-intercept at 4. The actual coordinates of the point are not important—only its location, which is shown, need concern us.

The arrows on each end of the parabola indicate that the curve continues upward.

## EXAMPLE 3

Graph:  $y = -x^2 + 2x + 3$.

Solution:   1.   Write in standard form and compare:

$$y = -x^2 + 2x + 3$$
$$y = ax^2 + bx + c$$

Thus $a = -1$, $b = 2$, $c = 3$.

Graph will be a parabolic curve.

2.   Since $a$ is negative, the curve will open down: ⌢.

3.   The $y$-intercept is $c = 3$.

4.   Find the $x$-intercepts, at most two, by solving:

$$0 = -x^2 + 2x + 3$$
$$0 = x^2 - 2x - 3$$
$$0 = (x - 3)(x + 1)$$
$$x = 3, \quad x = -1$$

The $x$-intercepts are 3 and −1.

5.   Find the coordinates of the vertex:

$$x = \frac{-b}{2a} = \frac{-(2)}{2(-1)} = \frac{-2}{-2} = 1$$

$$y = \frac{4ac - b^2}{4a} = \frac{4(-1)(3) - (2)^2}{4(-1)} = \frac{-12 - 4}{-4} = \frac{-16}{-4} = 4$$

The vertex is located at (1, 4).

6. Draw the axis of symmetry with a dashed line. It is a vertical line through the vertex, (1, 4).

7. Locate one point on either side of the vertex if necessary. In this example step 7 can be omitted because we already have points on either side of the vertex, the $y$-intercept, its matching point, and the two $x$-intercepts.

8. Plot the above points and connect them with a smooth, $\wedge$-shaped curve.

Answer:

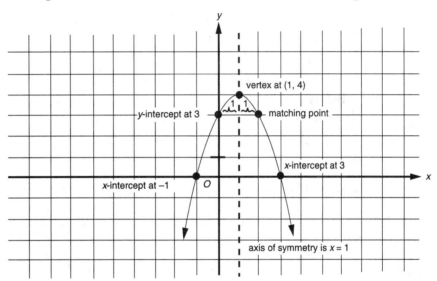

I'll do one more example before asking you to try to graph one. The next example also illustrates an alternative method for finding the $y$-coordinate of the vertex.

## EXAMPLE 4

Graph: $y = -x^2 - 1$.

Solution: 1. Write in standard form and compare:
$$y = -x^2 \qquad - 1$$
$$y = ax^2 + bx + c$$
Thus $a = -1$, $b = 0$, $c = -1$.
Graph will be a parabolic curve.

2. Since $a$ is negative, the curve will open down: $\wedge$.

3. The $y$-intercept is $c = -1$.

4. Find the $x$-intercepts, at most two, by solving:

$$0 = -x^2 - 1$$
$$x^2 = -1 \text{ has no solution}$$

Therefore there are no $x$-intercepts.

5. Find the coordinates of the vertex:

$$x = \frac{-b}{2a} = \frac{0}{2(-1)} = 0$$

To find the $y$-coordinate for the vertex, we can, of course, use the formula $(4ac - b^2)/4a$, but there is an alternative method.

Once the $x$-coordinate of the vertex is determined, the $y$-coordinate can be calculated by substituting $x$ into the original equation:

if $x = 0$, then $y = -(0)^2 - 1 = -1$.

The vertex is located at $(0, -1)$.

For this particular example, using the original equation was an easier way to determine $y$ than using the formula for the vertex, but the choice is yours.

6. Draw the axis of symmetry.
7. Locate one point on either side of the vertex.

let $x = 1$; then $y = -(1)^2 - 1 = -1 - 1 = -2$

let $x = -1$; then $y = -(-1)^2 - 1 = -1 - 1 = -2$

8. Plot the above points and connect them with a smooth, ∧-shaped curve.

Answer:

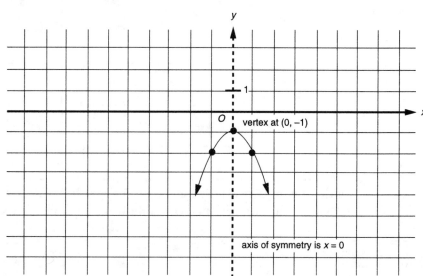

If more detail is needed, plot additional points.

_____

Now it is time for you to try a few problems.

_____

## Problem 1

Graph: $y = -x^2 + 4x + 5$.

_____

Solution: 1. Write in standard form and compare:

Thus $a =$ _____, $b =$ _____, and $c =$ _____.

2. The curve will open _____.
3. The $y$-intercept is _____.

4.  Find the $x$-intercepts, at most two, by solving:

    $0 = -x^2 + 4x + 5$

    $0 = x^2 - 4x - 5$

    Therefore the $x$-intercepts are _____ and _____ .

5.  Find the coordinates of the vertex:

    $x = \dfrac{-b}{2a} =$

    $y =$

6.  Draw the axis of symmetry.

7.  Locate one point on either side of the vertex if necessary.

    Step 7 can be omitted because we already have points on either side of the vertex, the $y$-intercept, its matching point, and the two $x$-intercepts.

8.  Plot the above points on the grid provided and connect them with a smooth curve.

Answer:

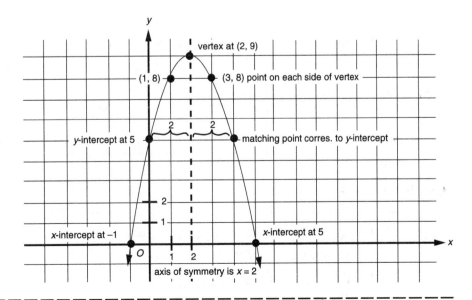

## Problem 2

Graph: $y = x^2 + 2$.

Solution:

Answer:

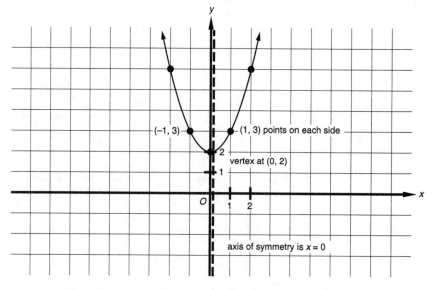

The next example illustrates still another method for locating the vertex of the parabola.

## EXAMPLE 5

Graph:   $y = 2x^2 - 4x$.

Solution:   1.   This is a quadratic equation with $a = 2$, $b = -4$, and $c = 0$.
2.   Parabola opens up.
3.   The $y$-intercept is 0.
4.   The $x$-intercepts are:

$$0 = 2x^2 - 4x$$
$$0 = 2x(x - 2)$$
$$x = 0, \qquad x = 2$$

5.   As before, we could locate the vertex using the formula, but in this example there is an easier method. Recall that the axis of symmetry is a vertical line through the vertex that cuts the curve in the middle; the two sides of the curve must match. Since in this example the $x$-intercepts are at 0 and 2, where must the axis of symmetry be? It must be midway between the two points. Thus the $x$-coordinate of the vertex is 1, the value in the middle of 0 and 2.

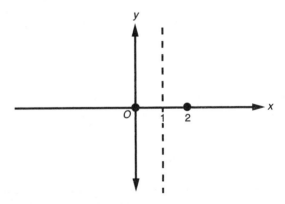

The $y$-coordinate may be calculated by either method discussed earlier. In this case I will substitute into the original equation:

if $x = 1$, $y = 2(1)^2 - 4(1) = 2 - 4 = -2$.

The vertex is located at $(1, -2)$.

Step 7 can be omitted because we already have points on either side of the vertex, the two $x$-intercepts.

Answer:

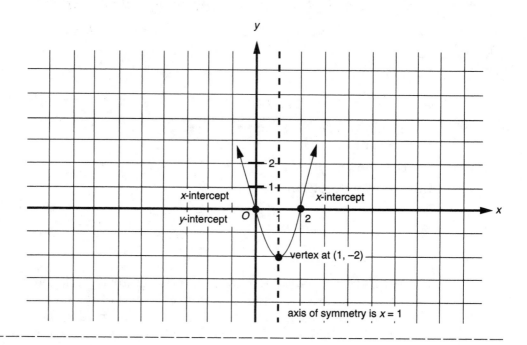

As you probably suspected, all quadratic equations are not as easy to work with as the preceding ones. Consider the next example.

## EXAMPLE 6

Graph: $y = x^2 + x - 1$.

Solution:  1.  This is a quadratic equation with $a = 1$, $b = 1$, and $c = -1$.
2.  Parabola opens up.
3.  The $y$-intercept is $-1$.
4.  Find the $x$-intercepts, at most two, by solving:

$$0 = x^2 + x - 1$$

Since the right-hand side is not easily factored, we will go on to step 5.

Note:  If the $x$-intercepts were required, the quadratic formula could be used to find them since we are unable to factor.

$$x = \frac{-b \pm \sqrt{b^2 - 4ac}}{2a}$$

$$= \frac{-1 \pm \sqrt{(1)^2 - 4(1)(-1)}}{2(1)}$$

$$= \frac{-1 \pm \sqrt{5}}{2} \text{ are the } x\text{-intercepts}$$

If you have a calculator, you will find that $\sqrt{5} \approx 2.24$, so

$$x = \frac{-1 + 2.24}{2} \quad \text{and} \quad x = \frac{-1 - 2.24}{2}$$
$$= 0.62 \qquad\qquad = -1.62$$

The $x$-intercepts are approximately 0.62 and $-1.62$.

5. The vertex is located at $\left( \dfrac{-1}{2}, \dfrac{-5}{4} \right)$ because:

$$x = \frac{-b}{2a} = \frac{-1}{2(1)} = \frac{-1}{2}$$

$$y = \frac{4ac - b^2}{4a} = \frac{4(1)(-1) - (1)^2}{4(1)} = \frac{-4 - 1}{4} = \frac{-5}{4}$$

6. Draw the axis of symmetry.

7. Locate one point on either side of the vertex:

let $x = -1$; then $y = (-1)^2 + (-1) - 1 = -1$

let $x = 0$; then $y = -1$

Answer:

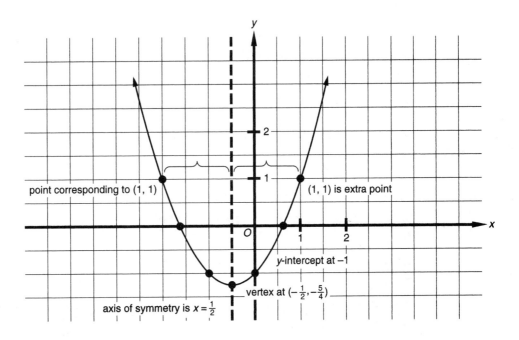

Comments: I approximated the $x$-intercepts. To get a better sense of where the curve actually crossed the $x$-axis, I plotted a convenient extra point, (1, 1), in quadrant I. There would be a corresponding point in quadrant II. Then, by connecting the five points with a smooth curve, the location of the $x$-intercepts is well determined.

# ALGEBRA AND THE CALCULATOR (Optional)

## Graphing Quadratic Equations

The procedure for graphing a quadratic equation and finding $x$-intercepts using a calculator is basically the same as for a linear equation. Finding the vertex can be done several different ways. I will explain my preference, which uses the steps and keystrokes introduced in the previous unit.

## EXAMPLE 7

Graph: $y = -4x^2 + 6x + 3$

      a.  Find the coordinates of the vertex.

      b.  Find the $x$-intercepts.

Solution:  Use the standard viewing window $\boxed{\text{ZOOM}}$ $\boxed{6}$.

Create a table of values for the equation with the following keystrokes:

$\boxed{\text{Y=}}$ $\boxed{\text{CLEAR}}$ $\boxed{(-)}$ $\boxed{4}$ $\boxed{\text{X,T,}\theta\text{,n}}$ $\boxed{x^2}$ $\boxed{+}$ $\boxed{6}$ $\boxed{\text{X,T,}\theta\text{,n}}$ $\boxed{+}$ $\boxed{3}$ $\boxed{\text{ENTER}}$.

To see the $y$-values, press $\boxed{\text{2nd}}$ $\boxed{\text{TABLE}}$.
To see the graph, press $\boxed{\text{GRAPH}}$.

    a.  Return to the table of $y$-values; press $\boxed{\text{2nd}}$ $\boxed{\text{TABLE}}$.
       This is a quadratic equation with $a = -4$, $b = 6$, and $c = 3$.

       The $x$-coordinate of the vertex is given by the formula $y = \dfrac{-b}{2a}$.

       Using the formula, key in the $x$-coordinate of the vertex as follows:

       $\boxed{(-)}$ $\boxed{6}$ $\boxed{\div}$ $\boxed{(}$ $\boxed{2}$ $\boxed{\times}$ $\boxed{(-)}$ $\boxed{4}$ $\boxed{)}$ $\boxed{\text{ENTER}}$.

   Note:  The entire denominator must be enclosed in parentheses so that $-6$ is divided by the product of 2 and $-4$, not just the first 2.

Answer:  The vertex is located at the point $(-0.75, 5.25)$.

As a check, use $\boxed{\text{TRACE}}$ to verify that the answer is reasonable.

    b.  To find the negative $x$-intercept, also called a zero, the steps are the same as those introduced in Unit 23.

       Press $\boxed{\text{2nd}}$ $\boxed{\text{CALC}}$ $\boxed{2}$.

       At the <u>Left Bound?</u> prompt, move the cursor to a point just to the left of the $x$-intercept and press $\boxed{\text{ENTER}}$.

       At the <u>Right Bound?</u> prompt, move the cursor to a point just to the right of the $x$-intercept and press $\boxed{\text{ENTER}}$.

       At the <u>Guess?</u> prompt, move the cursor between the boundaries and press $\boxed{\text{ENTER}}$.

Answer:  The negative $x$-intercept is $-0.396$.
       For practice, you find the positive $x$-intercept.

Answer:  The positive $x$-intercept is $1.896$.

To summarize, in Unit 25 we defined (and you should now be able to identify) a second-degree equation in two variables, of the type $y = ax^2 + bx + c$ with $a \neq 0$. Also, you should be able to graph such an equation, using a minimum number of well-chosen points, based on your knowledge of quadratic equations.

Recall that, in short, the graphing approach is as follows:

1. Write the equation in standard form and compare.

2. Determine whether the parabola opens up or down.

3. Find they $y$-intercept at $c$.

4. Solve $0 = ax^2 + bx + c$ to find the $x$-intercepts, if the equation is easily factored.

5. Find the coordinates of the vertex:

$$\left( \frac{-b}{2a}, \frac{4ac - b^2}{4a} \right)$$

6. Draw the axis of symmetry with a dashed line.

7. Locate a point on either side of the vertex if needed.

8. Plot the points, and connect them with a smooth, $\smile$-shaped curve.

Now try the following exercises.

## EXERCISES

Graph each of the following by using your knowledge of quadratic equations in two variables to plot a few well-chosen points. Optional—verify the reasonableness of your graph by using a graphing calculator.

1. $y = x^2 - 2x - 8$

2. $y = x^2 + 2x - 8$

3. $y = -4x^2$

4. $y = -x^2 - 4x - 3$

5. $y = x^2 - 6x + 5$

6. $y = 4 + 2x^2$

7. $y = x^2 - 4x + 4$

8. $y = 25 - x^2$

9. $y = x^2 + 6x + 9$

10. $y = 2x^2 + 4x - 1$

11. $y = 5x^2 - 20x + 11$

12. $y = -x^2 + 10x$

13. $y = 3x^2 - 3x + 2$

14. $y = 2x^2 - 12x + 3$

15. $y = \frac{-1}{4}x^2 + 3x - 8$

# UNIT 25

## Functions

This unit introduces the concept of a function. When you have completed the unit, you will be able to read and to use functional notation and identify the domain.

We will start with a diagram of a machine.

We can define our machine to do almost anything. Since this is a book about algebra, our machine is one that does calculations with numbers. Suppose that the numbers 5, 7, 8, and 9 are located as shown and that the machine "takes a number and adds 3."

It should be obvious what happens as each number is put into the machine.

If 5 is entered, the machine takes the number and adds 3, and out comes 8.

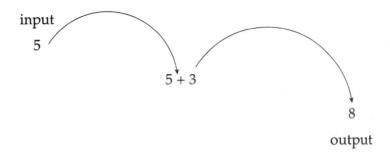

In a similar fashion, if 7 had been put in, the output would have been 10. If 8 had been put in, the output would have been 11. If 9 had been put in, the output would have been 12. Write the numbers 8, 10, 11, and 12 in the box in the machine diagram.

The values that can be put into the machine are referred to as the domain. The **domain** is the set of all possible values that can be used as input. In this example, the domain is the set of numbers 5, 7, 8, and 9, or $D = \{5, 7, 8, 9\}$. Only these four numbers can be put into this particular machine. Why? Because that is the way it was created.

The values that come out are referred to as the range. The **range** is the set of all possible values that are output. In the diagram the range is the set of numbers 8, 10, 11, and 12, or $R = \{8, 10, 11, 12\}$.

The description "takes the number and adds 3" is referred to as the **rule**.

Notice that our diagram has three parts: the domain, the range, and the rule.

The domain, often denoted by $x$, is the **independent variable**, whereas the range, often denoted as $y$, is the **dependent variable**. A way to remember the distinction between the two is that we can select any value from the domain to put into the machine, but once a specific value of $x$ is selected, such as 7, the output, or the $y$-value, is dependent on it.

By using $x$ to denote the independent variable, it is possible to rewrite the rule algebraically as $x + 3$. At the same time, we will give the machine the name $f$. Typically $f$ is the most common name used, although $g$ and $h$ are popular too.

The diagram has been drawn again to include all the terminology.

> Definition:   A **function** is a rule that assigns to each element in the domain one and only one element in the range.

Like the diagram, the definition refers to three parts: domain, range, and rule. The most important part of the definition, though, is the *one and only one* phrase. For each value put in, only a single value can come out if the expression is to be a function. Otherwise the expression is classified as a **relation**. Functions will be the major emphasis in this unit.

Obviously we don't want to have to draw a diagram every time we talk about a function. All of the above information can be combined into a single statement using functional notation.

$$y = f(x) = x + 3 \text{ with } D = \{5, 7, 8, 9\}$$

dependent     name   independent        rule       domain
variable             variable

The above statement is read "y equals f of x equals x plus 3, with domain D equal to the set of numbers 5, 7, 8, and 9.

Caution: $f(x)$ is read "f of x" and does not mean multiplication. It is used to indicate that the name of the function is f and that the variable inside the parentheses, in this example, $x$, is the independent variable.

It is time to stop and try some questions. Using $f$ as defined by the diagram, answer the following questions.

1.   If $x = 7$, what is $y$?          Answer:   10
2.   If $x = 9$, what is $y$?          Answer:   12
3.   If $y = 11$, what is $x$?         Answer:    8

I'm confident that you were able to answer each of these questions correctly without any difficulty whatsoever. The difficulty occurs when the same questions are asked using functional notation. Here are some examples.

## EXAMPLE 1

Continue to use $f$ as defined in the diagram, and find each of the following:

a.   $f(7)$.

b.   $f(9)$.

c.   $f^{-1}(11)$.

Solution:   a.   $f(7)$, read "f of 7," is asking, If $x = 7$, what is $y$? It is exactly the same as question 1 above, and the answer is 10, or $f(7) = 10$.

b.   $f(9)$ is asking the same thing as question 2 above but uses functional notation. The question is, If $x = 9$ and the rule is $f$, what is $y$? As in question 2 above, the answer is 12, or $f(9) = 12$.

c.   $f^{-1}(11)$ is read "f inverse of 11." The question is, If the $y$-value is 11, what is $x$? As in question 3 above, the answer is 8, or $f^{-1}(11) = 8$.

Here are a few problems for you to try. Use $f$ as defined in the diagram to answer each of the following.

## Problem 1

Find $f(5)$, $f^{-1}(10)$, and $f(8)$.

Solution:

Answers:   8; 7; 11

## Problem 2

Find $f(6)$. Be careful; the answer is not 9.

Solution:

Answer:   $f(6)$ is undefined because 6 is not an element of the domain. $f$ is defined only for the numbers 5, 7, 8, and 9, which was an arbitrary decision on my part; but that was the way I defined the machine.

## EXAMPLE 2

Given:   $y = g(x) = 2x + 5$ with the domain the set of real numbers.

Identify each part of the expression and

find:   $g(3)$, $g(-1)$, $g(0)$, $g(7)$, and $g(-5)$.

Solution:   $y$ is the dependent variable.

$x$ is the independent variable.

$g$ is the name of the function.

The rule is 2 times the number plus 5.

$g(x) = 2x + 5$
$g(3) = 2(3) + 5 = 6 + 5 = 11$
$g(-1) = 2(-1) + 5 = -2 + 5 = 3$
$g(0) = 2(0) + 5 = 0 + 5 = 5$
$g(7) = 2(7) + 5 = 14 + 5 = 19$
$g(-5) = 2(-5) + 5 = -10 + 5 = -5$

Although functions are commonly expressed in algebraic form, sets and graphs also can be used, as the next examples illustrate.

## EXAMPLE 3

Given:   $F = \{(1, 2), (3, 7), (3, 15)\}$.

Find:   a.   The domain.

b.   The range.

c.   $F(1)$.

d.   $F^{-1}(15)$.

e.   Is $F$ a function?

Solution:   $F$ is the name. The ordered pair (1, 2) means that, when $x = 1$, $y = 2$. Similarly, (3, 7) means that when $x = 3$, $y = 7$, and (3, 15) means that when $x = 3$, $y = 15$. In this example the actual rule is not given.

a.   The domain is the set of all $x$-values: $D = \{1, 3\}$. Note: There is no need to write the 3 twice.

b.   The range is the set of $y$-values: $R = \{2, 7, 15\}$.

c.   $F(1) = 2$. The notation is asking, When $x = 1$, what is $y$?

d.   $F^{-1}(15) = 3$. The notation is asking, When $y = 15$, what is $x$?

e.   No, $F$ is not a function. When $x = 3$, there are two different values for $y$, 7 and 15.

## EXAMPLE 4

Does the graph at the right specify a function?

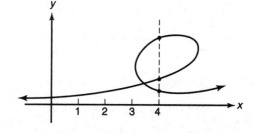

Solution:   No, for example, when $x = 4$, there are three values for $y$.

Notice that if you are able to draw a vertical line that intersects a graph at more than one point, the graph does not represent a function. This is referred to as the **vertical line test**.

When a function is expressed in algebraic form, the domain is seldom mentioned. Unless specified, the **domain** of a function is the set of all values for which the dependent variable is defined and real. This is usually the set of real numbers. Remember, the real numbers are all the counting numbers, zero, whole numbers, positive and negative integers, fractions, and rational and irrational numbers. There are two exceptions, or limitations, for the domain, but they are beyond the scope of this book.

# ALGEBRA AND THE CALCULATOR (Optional)

### Creating a Table of Values for a Function

In previous units we created tables of values for linear equations and quadratic equations. In this section the notation and terminology have been changed to create a table of values for a function, but the steps remain identical.

## EXAMPLE 5

Given:   $y = f(x) = 3.2x + 2.1$.

Find:   $f(14)$, $f(7.5)$, $f(1.45)$, $f(0)$, $f(-0.023)$, $f(-5)$, and $f(-8.1)$.

---

Solution:   This example is the same as the ones we did in Unit 23 except that now we are using functional notation and our example is called a linear function.

To enter the function, press the following keys.

Y= | CLEAR | 3 | · | 2 | X,T,θ,n | + | 2 | · | 1 | ENTER .

You should see the following:

To find the desired $y$-values,

press 2nd TABLE .

To find $f(14)$, press 1 4 ENTER .

To find $f(7.5)$, press 7 · 5 ENTER .

Repeat, entering the remaining values: 1.45, 0, −0.023, −5, and −8.1.

You should see the following:

| X | Y₁ | |
|---|---|---|
| 14 | 46.9 | |
| 7.5 | 26.1 | |
| 1.45 | 6.74 | |
| 0 | 2.1 | |
| −.023 | 2.0264 | |
| −5 | −13.9 | |
| −8.1 | −23.82 | |
| X=−8.1 | | |

Confirm that the function's graph is a line by viewing its graph.

### Changing the Dimensions of the Viewing Window

Thus far, most of the examples have been selected so that the graph could be seen using the standard viewing window ZOOM 6 . Recall that its dimensions are [−10, 10] by [−10, 10], which can be seen by pressing WINDOW .

In Unit 23 an approach for enlarging the viewing window for a linear equation was introduced. Another way to change the viewing window is to use the TRACE and ZOOM keys. This approach is illustrated in the next example.

## EXAMPLE 6

Graph   $y = f(x) = -1.2x^2 - 4.5x + 7$.

Solution:   In the previous unit, we explored quadratic equations such as the one in this example: $y = f(x) = -1.2x^2 - 4.5x + 7$. Now with the introduction of functional notation, this example can be referred to as a quadratic equation or a quadratic function. Using either terminology, the basic facts remain the same: Its graph will be a parabola opening down.

Use the standard viewing window $\boxed{\text{ZOOM}}$ $\boxed{6}$.

Create a table of values for the function with the following keystrokes:

$\boxed{\text{Y=}}$ $\boxed{\text{CLEAR}}$ $\boxed{(-)}$ $\boxed{1}$ $\boxed{.}$ $\boxed{2}$ $\boxed{\text{X,T,θ,n}}$ $\boxed{x^2}$ $\boxed{-}$ $\boxed{4}$ $\boxed{.}$ $\boxed{5}$ $\boxed{\text{X,T,θ,n}}$ $\boxed{+}$ $\boxed{7}$ $\boxed{\text{ENTER}}$ .

To see the $y$-values, press $\boxed{\text{2nd}}$ $\boxed{\text{TABLE}}$ .

To see the graph, press $\boxed{\text{GRAPH}}$ . This time you see only part of the parabola in the standard viewing window. The vertex is outside the viewing window.

Press $\boxed{\text{TRACE}}$ , and the value of the $y$-intercept, which is 7, appears at the bottom of the screen. Press $\boxed{\text{ZOOM}}$ $\boxed{3}$ $\boxed{\text{ENTER}}$ . More of the graph is visible; but it is difficult to see much detail.

Press $\boxed{\text{TRACE}}$ and move the cursor along using the $\boxed{\blacktriangleright}$ $\boxed{\blacktriangleleft}$ keys while noting the various $x$- and $y$-values at the bottom of the screen. These values provide clues as to what the new dimensions for the viewing window should be.

It appears that dimensions of [−10, 10] by [−10, 15] should be sufficient.

Press $\boxed{\text{WINDOW}}$ , change dimensions, and view the graph.

Answer:

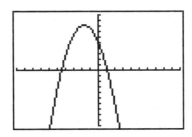

Before leaving this example completely, press $\boxed{\text{TRACE}}$ and move the cursor to some point close to one of the $x$-intercepts. Press $\boxed{\text{ZOOM}}$ $\boxed{2}$ $\boxed{\text{ENTER}}$ and notice what happens. The viewing window changes to magnify the portion of the graph centered around the position of the cursor. You can get as close to a point as desired by repeating the process of pressing $\boxed{\text{TRACE}}$ with $\boxed{\blacktriangleright}$ or $\boxed{\blacktriangleleft}$ to move the cursor and then pressing $\boxed{\text{ZOOM}}$ $\boxed{2}$ $\boxed{\text{ENTER}}$ .

I suggest you do some experimenting with all these features. Set various viewing window dimensions and practice with 2:ZOOM In, 3:ZOOM Out, and TRACE. Determine which options are best for the way you approach graphing. I have favorites, and I expect you will too.

You should now understand the basic concept of a function along with the terms *domain*, *range*, and *rule*. You should also be able to read and evaluate questions written in functional notation.

Before beginning the next unit you should do all the following exercises.

## EXERCISES

Given: $f(x) = x - 2$, find each of the following:

1. $f(7)$

2. $f(-3)$

3. $f(3.7)$

4. $f(0)$

Given: $g(x) = 3x$, find each of the following:

5. $g(2)$

6. $g(-5)$

7. $g(0)$

8. $g(-4.1)$

Given: $h(x) = 5x + 1$, find each of the following:

9. $h(10)$

10. $h(3)$

11. $h(0)$

12. $h(-4)$

Given: $f(x) = 1 - x^2$, find each of the following:

13. $f(3)$

14. $f(-5)$

15. $f(-1)$

Given: $G = \{(5, 6), (7, 7), (8, 5), (6, 11)\}$

16. State the domain for $G$.

17. State the range for $G$.

18. Find $G(8)$.

19. Find $G(5)$.

20. Find $G^{-1}(11)$.

21. Find $G^{-1}(5)$.

22. Is $G$ a function?

C23. Given the function $y = f(x) = 2x^2 - 1.5x + 2$, find $f(2.5)$ and $f(0.25)$.

C24. Given the function $y = f(x) = 3x^4 + 2x^3 - x^2 + 1$, find $f(2)$ and $f(1.5)$.

_____

# UNIT 26

## Solving Systems of Equations

The purpose of this unit is to provide you with an understanding of systems of equations. When you have finished the unit, you will be able to solve systems of two linear equations in two variables.

Recall the following definition from Unit 23:

---

Definition: An equation that has the form $ax + by = c$, with $a$, $b$, and $c$ being real numbers, $a$ and $b$ not both zero, is a **linear equation in two variables**.

---

By a "solution to an equation in two variables" we mean the ordered pairs of values of $x$ and $y$ that satisfy the equation. The procedure outlined earlier for locating points on the graph of an equation also yields solutions to the equation. In fact, solutions are often written as ordered pairs.

There are infinitely many solutions to an equation in two variables.

---

## EXAMPLE 1

Solve:  $x + y = 8$.

---

Solution:  Recall how to locate a point:

1. Select a convenient value for $x$.  Let $x = 2$.

2. Substitute into the equation.  $(2) + y = 8$

3. Solve for $y$.  $y = 6$

Then (2, 6) is a point on the graph of $x + y = 8$

and

$x = 2$ and $y = 6$ is called a solution to the equation $x + y = 8$.

Without going through the calculations, (3, 5) is a point on the graph of $x + y = 8$

and

$x = 3$ and $y = 5$ is another solution to the equation.

There are infinitely many points on the graph

and

there are infinitely many solutions to the equation, some of which are

$$x = 1 \quad \text{and} \quad y = 7 \quad \text{or simply (1, 7)}$$
$$x = 0 \quad \text{and} \quad y = 8 \quad \text{or} \qquad \text{(0, 8)}$$
$$x = 2.5 \quad \text{and} \quad y = 5.5 \quad \text{or} \qquad \text{(2.5, 5.5)}$$
$$x = -3 \quad \text{and} \quad y = 11 \quad \text{or} \qquad \text{(–3, 11)}$$

Find at least five more solutions yourself.

---

**A system of equations** means that there is more than one equation.

---

Definition:  **The solutions** to a system of equations are the pairs of values of $x$ and $y$ that satisfy *all* the equations in the system.

---

In this unit we will deal only with linear equations in two variables, saving quadratic equations for the next unit.

In general, the number of solutions to any system of *linear* equations is either one, none, or infinitely many.

$$\begin{cases} 2x + y = 24 \\ x - y = 6 \end{cases}$$

is an example of a system of linear equations. The solution to this system is $x = 10$ and $y = 4$, or simply (10, 4), because this pair of values satisfies *both* equations in the system.

To verify, check by substitution.

$$2x + y = 24 \qquad x - y = 6$$
$$2(10) + 4 \stackrel{?}{=} 24 \qquad 10 - 4 \stackrel{?}{=} 6$$
$$24 = 24 \qquad 6 = 6$$

The rest of the unit deals with finding such solutions.

There are numerous ways to solve systems of equations in two variables. The one developed in this unit is called **elimination by addition**. A variation of it, called **elimination by substitution**, will be explained toward the end of the unit.

## ELIMINATION BY ADDITION

A linear equation written in the form $ax + by = c$ is said to be in **standard form**. It is advisable to simplify and write both equations in standard form before attempting to solve a system of linear equations by elimination by addition.

The procedure I will use involves four steps:

1. **Multiply (if necessary)** the equations by constants so that the coefficients of the $x$ or the $y$ variable are the negatives of one another.

2. **Add** the equations from step 1.

3. **Solve** the equation from step 2.

4. **Substitute** the answer from step 3 back into one of the original equations, and solve for the second variable.

Here are some examples that illustrate the use of these steps in solving a system of linear equations.

## EXAMPLE 2

Solve: $\begin{cases} x + y = 7 \\ 5x - 3y = 11 \end{cases}$.

---

Solution:  Multiply the first equation by 3 so that the $y$-coefficients are the negatives of one another.

1. **Multiply.**  $3(x + y = 7)$
$5x - 3y = 11$

2. **Add.**  $3x + 3y = 21$
$5x - 3y = 11$

3. **Solve.**  $8x \quad = 32$
$x \quad = 4$

4. **Substitute back** into one of the original equations. I usually try to select the simpler of the two original equations for the substitution.

If $x = 4$ and $x + y = 7$,
$$(4) + y = 7,$$
$$y = 3.$$

Answer: Solution to system is $x = 4$ and $y = 3$ or, written another way, $(4, 3)$.

Note: Back in step 1 we could just as well have multiplied the equation by $-5$ so that the $x$-coefficients would have been the negatives of one another.

---

## EXAMPLE 3

Solve: $\begin{cases} 3x + 2y = 12 \\ y = 2x - 1 \end{cases}$.

---

Solution: Write the equations in standard form.

$$3x + 2y = 12$$
$$-2x + y = -1$$

Multiply the second equation by $-2$ so that the $y$-coefficients are the negatives of one another.

1. **Multiply.** $\quad 3x + 2y = 12$

$\quad\quad\quad\quad -2(-2x + y = -1$

2. **Add.** $\quad\quad\quad 3x + 2y = 12$

$\quad\quad\quad\quad\quad 4x - 2y = \ 2$

3. **Solve.** $\quad\quad\quad 7x \quad\quad = 14$

$\quad\quad\quad\quad\quad\quad x \quad\quad = 2$

4. **Substitute back** into an original equation.

If $x = 2$ and $y = 2x - 1$,
$$y = 2(2) - 1,$$
$$y = 3.$$

Answer: Solution to system is $x = 2$ and $y = 3$.

---

## EXAMPLE 4

Solve: $\begin{cases} 3x - y = -7 \\ 5y + 5 = -5x \end{cases}$.

---

Solution: Simplify the second equation, and write it in standard form before starting the procedure.

Rewrite. $5x + 5y = -5$

Divide by 5. $x + y = -1$

The equations are now: $3x - y = -7$

$x + y = -1$

1. **Multiply**—not necessary.

2. **Add.** $3x - y = -7$

$x + y = -1$

3. **Solve.** $4x \quad = -8$

$x \quad = -2$

4. **Substitute back** into an original equation.

If $x = -2$ and $x + y = -1$,

$(-2) + y = -1,$

$y = 1.$

Answer: Solution to system is $x = -2$ and $y = 1$.

_ _ _ _ _ _ _ _ _ _ _ _ _ _ _ _ _ _ _ _ _ _ _ _ _ _ _ _ _ _ _ _ _ _ _

From an algebraic point of view,

$x = -2$ and $y = 1$ is the solution to the system of equations $3x - y = -7$ and $x + y = -1$.

From a geometric point of view,

$(-2, 1)$ is the point of intersection for two lines whose equations are given by $3x - y = -7$ and $x + y = -1$.

To verify that $(-2, 1)$ is the point of intersection, graph the two lines.

---

## Problem 1

Graph: $\begin{cases} 3x - y = -7 \\ 5y + 5 = -5x \end{cases}$.

_ _ _ _ _ _ _ _ _ _ _ _ _ _ _ _ _ _ _ _ _ _ _ _ _ _ _ _ _ _ _ _ _ _ _

Solution:

Remember:    To graph a line, create a table of values by solving the equation for $y$. Next select three convenient values for $x$ and solve for $y$ by substitution. Plot the three points and connect them with a straight line.

Answer:

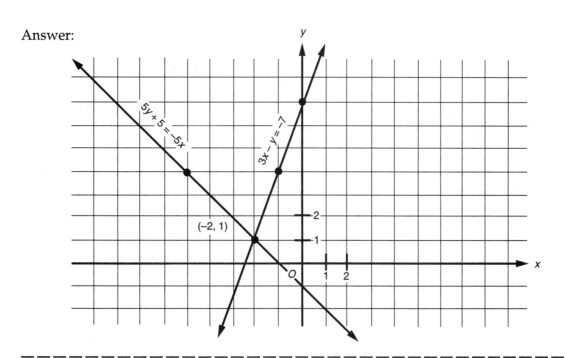

Now it is your turn to solve a system of equations.

## Problem 2

Solve:    $\begin{cases} 4x - y - 1 = 0 \\ 2x = 17 - y \end{cases}$.

Solution:

Hint:   First write equations in standard form.

      1.  **Multiply**, if necessary.

      2.  **Add.**

      3.  **Solve.**

      4.  **Substitute back** into an original equation.

Answer:   Solution to system is $x = 3$ and $y = 11$.

- - - - - - - - - - - - - - - - - - - - - - - - - - - - - - - - - - -

Verify, by doing Problem 3, that your answer to Problem 2 is the intersection of the two lines.

___

## Problem 3

Graph:   $\begin{cases} 4x - y - 1 = 0 \\ 2x = 17 - y \end{cases}$ .

- - - - - - - - - - - - - - - - - - - - - - - - - - - - - - - - - - -

Solution:

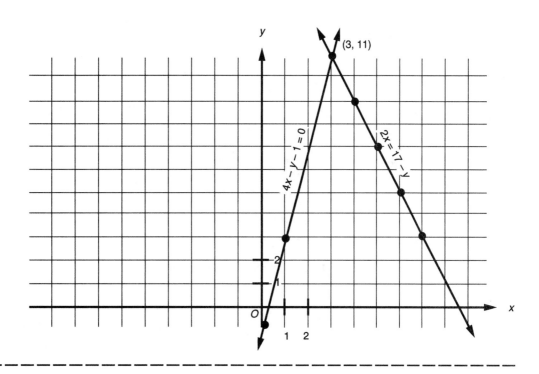

Answer:

_____

## Problem 4

Solve: $\begin{cases} x + 2y = 1 \\ 5x + 3y = 26 \end{cases}$.

_____

Solution:  1.  **Multiply.**

2.  **Add.**

3.  **Solve.**

4.  **Substitute back** into an original equation.

Answer:  Solution to system is $x = 7$ and $y = -3$.

_____

Many times it is necessary to multiply the two equations by different constants in order to make the coefficients of the $x$ or the $y$ variable be the negatives of one another.

═══════════════════════════════════════════════════

## EXAMPLE 5

Solve:  $\begin{cases} 2x + 3y = 9 \\ 5x + 2y = 17 \end{cases}$.

_____

Solution:  1.  **Multiply.**  $-5(2x + 3y = 9)$

$2(5x + 2y = 17)$

2.  **Add.**  $-10x - 15y = -45$

$\underline{10x + 4y = 34}$

3.  **Solve.**  $-11y = -11$

$y = 1$

4.  **Substitute back** into an original equation.

If $y = 1$ and $2x + 3y = 9$,

$2x + 3(1) = 9,$

$2x = 6,$

$x = 3.$

Answer:  Solution to system is $x = 3$ and $y = 1$.

Did you notice that there were several possibilities for the constants used in step 1? We could just as well have multiplied the first equation by 2 and the second equation by –3.

What other combinations could have been used?

_____

Here are two problems for you to solve on your own.

## Problem 5

Solve: $\begin{cases} 3x - 2y = 5 \\ -4x + 3y = 1 \end{cases}$.

Solution:

Answer: $x = 17$ and $y = 23$

As you probably suspected, all systems do not have nice integral values as solutions. Be prepared for fractions.

## Problem 6

Solve: $\begin{cases} \dfrac{x}{3} + y = 3 \\ 4x + 2y = 0 \end{cases}$.

Solution:

Hint: Simplify the first equation by clearing the fractions before attempting to solve this system of linear equations.

Answer: $x = \dfrac{-9}{5}$ and $y = \dfrac{18}{5}$

Recall that at the beginning of this unit it was stated that the number of solutions to a system of linear equations will be one, none, or infinitely many. Thus far, we have seen only systems with one solution. Examples 6 and 7 will examine the other two situations.

## EXAMPLE 6

Solve: $\begin{cases} 3x + 4y = 2 \\ 2y = 4 - \dfrac{3}{2}x \end{cases}.$

Solution: First simplify the second equation, and write it in standard form.

$$2\left(2y = 4 - \frac{3}{2}x\right)$$

$$4y = 8 - 3x$$

$$3x + 4y = 8$$

1. **Multiply.** $-1(3x + 4y = 2)$

    $3x + 4y = 8$

2. **Add.** $\quad -3x - 4y = -2$

    $\underline{3x + 4y = 8}$

    $0 = 6$

But $0 = 6$ is a false statement; and because $0 = 6$ is a false statement, there are no values of $x$ and $y$ that would ever satisfy this equation.

Answer: There is no solution to the system.

The equations are said to be **inconsistent** when there is no solution to the system.

Geometrically, no solution to a system of two linear equations means that there is no point of intersection for the two lines. The lines are parallel.

We will verify, by graphing, that the system in Example 6 represents two parallel lines. I will start Problem 7 but let you finish it.

## Problem 7

Graph: $\begin{cases} 3x + 4y = 2 \\ 2y = 4 - \dfrac{3}{2}x \end{cases}.$

Solution: Create a table of values by solving each equation for $y$.

$$3x + 4y = 2$$
$$4y = -3x + 2$$
$$y = \frac{-3}{4}x + \frac{2}{4}$$

$$2y = 4 - \frac{3}{2}x$$
$$4y = 8 - 3x$$
$$y = 2 - \frac{3}{4}x$$

Select three convenient values for $x$. Suggestions for the first line are to let $x = -6, -2, 2,$ or $6$. Suggestions for the second line are to let $x = -4, 0,$ and $4$.

Solve for $y$-values by substitution. Plot the three points for each line and connect them with a straight line.

Answer:

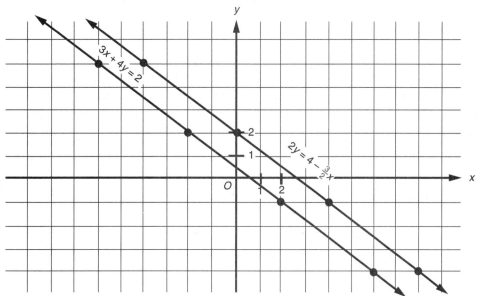

## EXAMPLE 7

Solve:   $\begin{cases} 3x - y = 5 \\ 6x - 2y - 10 = 0 \end{cases}$.

---

Solution:   $-2(3x - y = 5)$

$6x - 2y = 10$

$-6x + 2y = -10$

$6x - 2y = 10$

$0 = 0$     is a true statement.

This time our procedure results in a true statement or, as defined in an earlier unit, an identity.

All values of $x$ and $y$ would satisfy this equation, because it is an identity.

Answer:   There are infinitely many solutions to the system.

Every solution to the first equation will satisfy the second equation as well. A closer inspection of the two equations should reveal why this is true; the two equations are identical.

Find at least three solutions to the system.

Several possibilities include $(0, -5)$, $(2, 1)$, $(-2, -11)$.

---

The equations are said to be **dependent** when there are infinitely many solutions to the system.

Geometrically, "many solutions to a system of two linear equations" means there are infinitely many points of intersection for the two lines. The lines coincide.

Let's review what has just been stated.

---

**No solution to a system** is indicated when the elimination method leads to a false statement, such as $0 = 6$. The equations are said to be **inconsistent**. And the lines are **parallel**.

**Infinitely many solutions** to a system are indicated when the elimination method leads to an identity, such as $0 = 0$. The equations are said to be **dependent**. And the lines **coincide**.

---

## Problem 8

Solve:   $\begin{cases} 5x - y = 3 \\ 2y = 10x - 6 \end{cases}$.

---

Solution:

Answer:   There are infinitely many solutions.

---

## Problem 9

Solve: $\begin{cases} 3x + 4y = 11 \\ 2y = 4 - \dfrac{3}{2}x \end{cases}$.

---

Solution:

**Answer:**   There is no solution to the system.

# ELIMINATION BY SUBSTITUTION

Elimination by substitution is a variation of the method we have used thus far to solve systems of equations. If one of the variables has a coefficient of 1, it is sometimes more efficient to eliminate that variable by substitution rather than by multiplication and addition. Let me illustrate what I mean with the next example.

## EXAMPLE 8

Solve: $\begin{cases} y = 2x - 8 \\ 3x + 2y = 12 \end{cases}$.

Solution: Use $y = 2x - 8$ to substitute into the second equation.

$$3x + \phantom{0}2y \phantom{00} = 12$$
$$3x + 2(2x - 8) = 12 \qquad \text{Replace } y \text{ with } 2x - 8.$$
$$3x + \phantom{0}4x - 16 = 12$$
$$7x \phantom{000} = 28$$
$$x \phantom{0000} = 4$$

Finish the problem as before by substituting back into an original equation to find $y$:

$$y \phantom{000} = 0$$

Answer: Solution to system is $x = 4$ and $y = 0$.

I use elimination by substitution whenever possible, because it often requires less rewriting of the equations.

## EXAMPLE 9

Solve: $\begin{cases} y = 3x + 1 \\ 3x + 4y = -26 \end{cases}$.

Solution: Use $y = 3x + 1$ to substitute into the second equation.

$$3x + 4y \phantom{00} = -26$$
$$3x + 4(3x + 1) = -26 \qquad \text{Replace } y \text{ with } 3x + 1.$$
$$3x + 12x + 4 = -26$$
$$15x \phantom{000} = -30$$
$$x \phantom{0000} = -2$$

and, by substituting back into an original equation,

$$y \phantom{000} = -5$$

Answer: Solution to system is $x = -2$ and $y = -5$.

Let's do one more example, and then you can try some problems.

## EXAMPLE 10

Solve: $\begin{cases} 5x - 2y = 11 \\ y = \dfrac{5}{2}x - 3 \end{cases}$ .

Solution:  Use $y = \dfrac{5}{2}x - 3$ to substitute into the first equation.

$$5x - 2y = 11$$

$$5x - 2\left(\dfrac{5}{2}x - 3\right) = 11 \quad \text{Replace } y \text{ with } \dfrac{5}{2}x - 3.$$

$$5x - 5x + 6 = 11$$

$$6 = 11 \text{ is a false statement.}$$

Answer:  There is no solution to the system.

Here are two problems for you to solve. The first one is easy; the second is similar to the ones used as examples.

## Problem 10

Solve: $\begin{cases} 3x + 5y = 14 \\ y = -2 \end{cases}$ .

Solution:

Answer:  The solution to the system is $x = 8$ and $y = -2$.

## Problem 11

Solve by substitution: $\begin{cases} y = 7x + 2 \\ x - 3y = -6 \end{cases}$.

Solution:

Answer: $(0, 2)$

# ALGEBRA AND THE CALCULATOR (Optional)

### Finding Points of Intersection

We will revisit Example 9, but this time we will find the solution using a graphing calculator.

## EXAMPLE 11

Solve: $\begin{cases} y = 3x + 1 \\ 3x + 4y = -26 \end{cases}$.

Solution: Graph and find the point of intersection using a graphing calculator. First solve the second equation for $y$.

$$3x + 4y = -26$$

$$4y = -3x - 26$$

$$\frac{4y}{4} = \frac{-3x - 26}{4}$$

$$y = \frac{-3x - 26}{4}$$

Use the standard viewing window $\boxed{\text{ZOOM}}\ \boxed{6}$.

Key in *both* functions as follows:

$\backslash Y_1$ $\boxed{\text{Y=}}$ $\boxed{3}$ $\boxed{\text{X,T,}\theta\text{,n}}$ $\boxed{+}$ $\boxed{1}$ $\boxed{\text{ENTER}}$.

$\backslash Y_2$ $\boxed{\text{Y=}}$ $\boxed{(}$ $\boxed{(-)}$ $\boxed{3}$ $\boxed{\text{X,T,}\theta\text{,n}}$ $\boxed{-}$ $\boxed{2}$ $\boxed{6}$ $\boxed{)}$ $\boxed{\div}$ $\boxed{4}$ $\boxed{\text{ENTER}}$.

and view the graph.

To find the point of intersection, press $\boxed{\text{2nd}}$ $\boxed{\text{CALC}}$ $\boxed{5}$.

At the <u>First curve?</u> prompt, move the cursor along the first line to where the intersection appears to be and press ENTER.

At the <u>Second curve?</u> prompt, more the cursor along the second line to where

the intersection appears to be and press ENTER.

At the <u>Guess?</u> prompt, move the cursor between the boundaries shown on the screen by the two arrows, ▶ ◀, and press ENTER.

Answer:   Intersection   $X = -2$   $Y = -5$   appears at the bottom of the screen.

You should now be able to solve any system of two linear equations in two variables. Remember that elimination by addition is one method used to solve such systems; briefly stated, the four basic steps involved are:

1.  **Multiply**, if necessary.

2.  **Add.**

3.  **Solve.**

4.  **Substitute back.**

Also remember that the number of solutions to any system of linear equations is either one, none, or many.

Before beginning the next unit you should do the following exercises.

## EXERCISES

Solve these systems, using either elimination by addition or elimination by substitution:

1. $\begin{cases} 5x + 2y = 22 \\ 3x - 2y = 10 \end{cases}$

2. $\begin{cases} 13x - 5y = -5 \\ x + y = 1 \end{cases}$

3. $\begin{cases} -7x + y = 2 \\ 2x - y = 8 \end{cases}$

4. $\begin{cases} 5x + 3y = 1 \\ -2x + 5y = 12 \end{cases}$

C5. $\begin{cases} 4x - 3y = -10 \\ 5x + 2y = 22 \end{cases}$

6. $\begin{cases} 2x + y = 0 \\ x - y = 1 \end{cases}$

C7. $\begin{cases} y = 7x + 2 \\ x - 3y = -6 \end{cases}$

8. $\begin{cases} -2x + 17y = 6 \\ 11x - 5y = -33 \end{cases}$

9. $\begin{cases} 5x + 2y = 50 \\ 4x - 3y = -52 \end{cases}$

10. $\begin{cases} 4x + y = 13 \\ x - 3y = 0 \end{cases}$

C11. $\begin{cases} y = 2x + 3 \\ y + 2 = 4x + 1 \end{cases}$

12. $\begin{cases} x + 4y = 8 \\ y = -\dfrac{1}{4}x - 7 \end{cases}$

13. $\begin{cases} x - 3y = 2 \\ 4x - 10y = 10 \end{cases}$

14. $\begin{cases} 2x + \dfrac{1}{2}y = 2 \\ 6x - y = 1 \end{cases}$

15. $\begin{cases} x = 7 - \dfrac{1}{2}y \\ 8x - y = -4 \end{cases}$

16. $\begin{cases} y = -\dfrac{2}{5}x + 4 \\ x = \dfrac{1}{3}y - 7 \end{cases}$

17. $\begin{cases} 3x - 2y = 8 \\ -6x + 4y = 10 \end{cases}$

18. $\begin{cases} 11x - y = 10 \\ 2x + y = 7 \end{cases}$

19. $\begin{cases} 2a + 3b = 10 \\ 3a + 2b = 10 \end{cases}$

20. $\begin{cases} y = -\dfrac{2}{7}x - 1 \\ 2x = -7(y + 1) \end{cases}$

# UNIT 27

## Solving Systems of Equations (Continued)

In Unit 26 you learned to solve systems of linear equations. When you have finished this unit, you will be able to solve systems of equations containing one linear equation and one quadratic equation.

First, recall how we define a solution:

> Definition: The **solutions** to a system of equations are the ordered pairs of values of $x$ and $y$ that satisfy *all* the equations in the system.

As was previously stated, either elimination by addition or elimination by substitution may be used to solve linear systems of equations. When one of the equations is a quadratic, elimination by substitution is used.

## ELIMINATION BY SUBSTITUTION

When one equation of the system is a quadratic, the procedure I will use involves three steps:

1. Use the quadratic equation to **substitute** into the linear equation. If necessary, write the quadratic equation in standard form first.
2. **Solve** the equation from step 1 either by factoring or by using the quadratic formula.
3. **Substitute** the answer back into one of the original equations, and solve for the second variable.

The examples that follow illustrate the use of these steps in solving a system of equations containing one quadratic equation.

## EXAMPLE 1

Solve:  $\begin{cases} -3x + 2y = 2 \\ y = x^2 \end{cases}$ .

---

Solution:   Use $y = x^2$ to substitute into the linear equation.

$$-3x + 2y = 2$$

$$\downarrow$$

$-3x + 2(x^2) = 2$   Replace $y$ with $x^2$.

$-3x + 2x^2 = 2$   Solve the new equation.

$2x^2 - 3x - 2 = 0$

$(2x + 1)(x - 2) = 0$

$2x + 1 = 0$   or   $x - 2 = 0$

$x = -\dfrac{1}{2}$   or   $x = 2$

Substitute back into an original equation.

If $x = -\dfrac{1}{2}$              If $x = 2$

and $y = x^2$,               and $y = x^2$,

$y = \left(-\dfrac{1}{2}\right)^2$,               $y = (2)^2$,

$y = \dfrac{1}{4}.$               $y = 4.$

Answer:   There are two solutions to the system:

$$x = -\frac{1}{2} \text{ and } y = \frac{1}{4} \quad \left(-\frac{1}{2}, \frac{1}{4}\right)$$

and

$$x = 2 \text{ and } y = 4 \qquad (2, 4)$$

# GRAPHING THE SYSTEM WHEN ONE EQUATION IS A QUADRATIC

Recall that the graph of a linear equation is a line and the graph of a quadratic equation is a parabola.

From the algebraic point of view

the system:   $\begin{cases} -3x + 2y = 2 \\ y = x^2 \end{cases}$

has two solutions:   $x = -\dfrac{1}{2}$ and $y = \dfrac{1}{4}$

and

$x = 2$ and $y = 4$

From a geometric point of view

$\left( -\dfrac{1}{2}, \dfrac{1}{4} \right)$ and $(2, 4)$ are the points of intersection for the line and parabola whose

equations are given by $-3x + 2y = 2$ and $y = x^2$.

To verify that $\left( -\dfrac{1}{2}, \dfrac{1}{4} \right)$ and $(2, 4)$ are the points of intersection, graph the two equations

as instructed in Problem 1.

---

## Problem 1

Graph:   $\begin{cases} -3x + 2y = 2 \\ y = x^2 \end{cases}$.

---

Solution:

Answer:

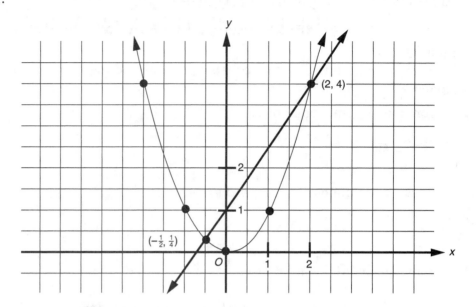

==========

# EXAMPLE 2

Solve:   $\begin{cases} y = -x^2 + 2x + 3 \\ y = 3 - x \end{cases}$ .

----------

Solution:   Use $y = -x^2 + 2x + 3$ to substitute into the equation.

$$y = 3 - x$$

$$\overbrace{-x^2 + 2x + 3} = 3 - x \qquad \text{Solve the new equation.}$$

$$-x^2 + 3x = 0$$

$$-x(x - 3) = 0$$

$$-x = 0 \qquad \text{or} \qquad x - 3 = 0$$

$$x = 0 \qquad\qquad\qquad x = 3$$

Substitute back into an original equation.

| If $x = 0$ | If $x = 3$ |
|---|---|
| and $y = 3 - x$, | and $y = 3 - x$, |
| $= 3 - 0$, | $= 3 - 3$, |
| $= 3$. | $= 0$. |

Answer:   There are two solutions to the system:

$$x = 0 \text{ and } y = 3 \qquad (0, 3)$$

and

$$x = 3 \text{ and } y = 0 \qquad (3, 0).$$

Are you ready to try solving one yourself?

## Problem 2

Solve: $\begin{cases} y = x^2 + x + 1 \\ y = -x \end{cases}$ .

Solution:

Answer:   $(-1, 1)$

Verify, by doing Problem 3, that your answer to Problem 2 is the intersection of the parabola and the line.

## Problem 3

Graph: $\begin{cases} y = x^2 + x + 1 \\ y = -x \end{cases}$ .

Solution:

Answer:

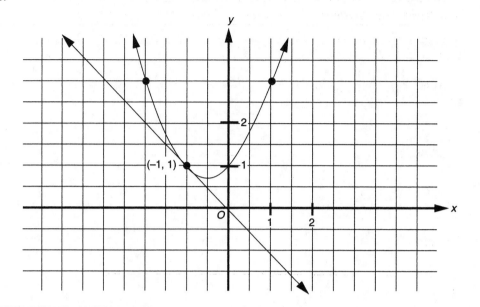

---

# EXAMPLE 3

Solve:   $\begin{cases} y = x^2 + 5x + 4 \\ 2x - 3y = 15 \end{cases}$.

---

Solution:   Use $y = x^2 + 5x + 4$ to substitute into the equation.

$$2x - \quad 3y \quad = 15$$

$$2x - \overbrace{3(x^2 + 5x + 4)} = 15$$

$$2x - 3x^2 - 15x - 12 = 15$$

$$-3x^2 - 13x - 27 = 0$$

Not readily factorable; use quadratic formula with $a = -3$, $b = -13$, and $c = -27$.

$$x = \frac{-b \pm \sqrt{b^2 - 4ac}}{2a}$$

$$= \frac{-(-13) \pm \sqrt{(-13)^2 - 4(-3)(-27)}}{2(-3)}$$

$$= \frac{13 \pm \sqrt{169 - 324}}{-6}$$

$$= \frac{13 + \sqrt{-155}}{-6}$$

No solution.

Answer:   There is no solution to the system.

Recall what was stated in Unit 26 regarding no solution to a system: The equations are said to be **inconsistent** when there is no solution to the system.

Geometrically, no solution to a system of equations means that there is no point of intersection for the two graphs.

Verify, by doing Problem 4, that the system in Example 3 represents a line and a parabola that do not intersect.

## Problem 4

Graph: $\begin{cases} y = x^2 + 5x + 4 \\ 2x - 3y = 15 \end{cases}$ .

Solution:

Answer:

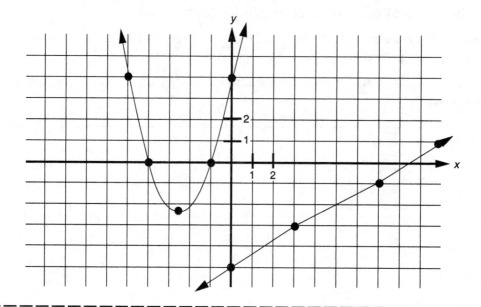

Here are a few more systems for you to solve. For additional practice, you might consider graphing them as well.

## Problem 5

Solve: $\begin{cases} y = -x^2 + 4x + 5 \\ x - y = -5 \end{cases}$.

Solution:

Answer:   The system has two solutions: (0, 5) and (3, 8).

## Problem 6

Solve:  $\begin{cases} y = x^2 + 2 \\ 3x + 4y = -8 \end{cases}$ .

Solution:

Answer:   There is no solution to the system.

Just for fun, try solving the next problem. It involves a system of equations, but this time both equations are quadratics. I have confidence that you can do the problem without any explanation on my part.

## Problem 7

Solve:  $\begin{cases} y = x^2 - 6x + 5 \\ y = x^2 - 4x + 4 \end{cases}$ .

Solution:

Answer:  $\left( \dfrac{1}{2}, 2\dfrac{1}{4} \right)$

# ALGEBRA AND THE CALCULATOR (Optional)

### Finding Points of Intersection

We will revisit Problem 2, but this time we will find the solution using a graphing calculator.

## EXAMPLE 4

Given: $\begin{cases} y = x^2 + x + 1 \\ y = -x \end{cases}$.

Graph and find the point of intersection.

Solution: Use the standard viewing window $\boxed{\text{ZOOM}}$ $\boxed{6}$.

Key in *both* equations as follows:

$\backslash Y_1$ $\boxed{\text{Y=}}$ $\boxed{\text{CLEAR}}$ $\boxed{\text{X,T,θ,n}}$ $\boxed{x^2}$ $\boxed{+}$ $\boxed{\text{X,T,θ,n}}$ $\boxed{+}$ $\boxed{1}$ $\boxed{\text{ENTER}}$.

$\backslash Y_2$ $\boxed{\text{Y=}}$ $\boxed{\text{CLEAR}}$ $\boxed{\text{(–)}}$ $\boxed{\text{X,T,θ,n}}$ $\boxed{\text{ENTER}}$.

and view the graph. The graph should match the one shown in Problem 3.

To find the point of intersection, press $\boxed{\text{2nd}}$ $\boxed{\text{CALC}}$ $\boxed{5}$.

At the <u>First curve?</u> prompt, move the cursor along the parabola to where the intersection appears to be and press $\boxed{\text{ENTER}}$.

At the <u>Second curve?</u> prompt, move the cursor along the line to where the intersection appears to be and press $\boxed{\text{ENTER}}$.

At the <u>Guess?</u> prompt, move the cursor between the boundaries shown on the screen by the two arrows, ▶ ◀, and press $\boxed{\text{ENTER}}$.

Answer: Intersection $X = -1$   $Y = 1$ appears at the bottom of the screen.

You should now be able to solve any system of equations containing one linear equation and one quadratic equation. Remember that elimination by substitution is the method used to solve such systems; briefly stated, the three basic steps involved are:

1.  **Substitute into** the linear equation.
2.  **Solve.**
3.  **Substitute back.**

Before beginning the next unit you should solve the following systems. Optional—verify your answers by reworking the problems on a graphing calculator before checking your answers with the back of the book.

## EXERCISES

Solve these systems using elimination by substitution.

1. $\begin{cases} y = 3x^2 - 7x + 11 \\ x = 2 \end{cases}$

2. $\begin{cases} y = 4x^2 \\ -4x + y = 0 \end{cases}$

3. $\begin{cases} y = x^2 + 5x \\ y = -2x + 44 \end{cases}$

4. $\begin{cases} y = -x^2 - 1 \\ -x + 5y = -5 \end{cases}$

5. $\begin{cases} y = x^2 + 5x - 21 \\ y = x \end{cases}$

6. $\begin{cases} y = x^2 + 2x + 5 \\ -4x + 2y = -7 \end{cases}$

7. $\begin{cases} y = 2x^2 - 4x \\ y = 8x - 18 \end{cases}$

8. $\begin{cases} y = 3x^2 + 8x - 2 \\ y = 1 \end{cases}$

9. $\begin{cases} y = x^2 + 13 \\ y = -10x - 12 \end{cases}$

10. $\begin{cases} y = x^2 + 5x + 1 \\ y = x^2 + 3x + 9 \end{cases}$   Here is another one for fun.

C11.   Given $y = x^2 - 5x + 2$ and $y = 5x - 8$, find the point of intersection in quadrant IV.

# UNIT 28

## Solving Inequalities—First-Degree

The purpose of this unit is to provide you with an understanding of inequalities. When you have finished the unit, you will be able to solve *all* first-degree inequalities.

### INEQUALITY SIGNS

There are two inequality signs:

> < read "less than"
> > read "greater than"

Sometimes an inequality sign and an equal sign are combined:

> ≤ read "less than or equal to"
> ≥ read "greater than or equal to"

For our purposes in this book we will use a "commonsense" definition of <, based on your familiarity with the number line. Remember the number line? It looks like this:

Number Line

Definition:   $a < b$ if $a$ is to the left of $b$ on the number line.

Here are some examples:

$2 < 3$ because 2 is to the left of 3 on the number line.

$-2 < 5$ because $-2$ is to the left of 5 on the number line.

$-5 < -2$ because $-5$ is to the left of $-2$.

$-7 < -6$ because $-7$ is to the left of $-6$.

Before proceeding, write a few more examples yourself.

The sign $>$ can be defined in a similar manner. How would the definition read? An alternative approach is to define $>$ as follows:

$b > a$ if and only if $a < b$.

In words, $b$ is greater than $a$ if $a$ is less than $b$.

For instance:

$3 > 2$ because $2 < 3$.

$0 > -1$ because $-1 < 0$.

$-4 > -5$ because $-5 < -4$.

**Notice that the inequality sign always opens to the larger number.**

Observe that $5 > 2$ has the same meaning as $2 < 5$. Both statements indicate that 5 is the larger number and 2 is the smaller number. Thus:

$3 > -1$ can be rewritten as $-1 < 3$.

$-8 < -2$ can be rewritten as $-2 > -8$.

$3 > x$ can be rewritten as $x < 3$.

$0 < y$ can be rewritten as $y > 0$.

I suggest rewriting most inequalities using $<$, because it is clearer to visualize the relationship of the numbers and the symbol, reading from left to right, on the number line.

The direction to which an inequality sign points is referred to as its **sense**. For example, $x < 3$ and $a < b$ are said to have the same sense because their symbols are pointing in the same direction. In contrast, $a > c$ and $2 < 5$ are said to be of opposite sense because their symbols are pointing in different directions.

# FIRST-DEGREE INEQUALITIES

Remember that in any first-degree equation in one variable:

There is only one variable.

The variable is involved in only the four fundamental arithmetic operations.

The variable is never multiplied by itself.

The variable is never in a denominator.

A **first-degree inequality** in one variable has the same characteristics as a first-degree equation except that in place of the equal sign there is an inequality sign.

Here are some examples of first-degree inequalities:

$$2x + 5 < 7$$
$$3(x - 1) + x \geq 2 - x$$
$$1 - x \leq 15(3 + 2x) - x$$

To solve a first-degree inequality, find the values of $x$ that satisfy the inequality. The basic strategy is the same as that used to solve first-degree equations—get all terms involving $x$ on one side of the inequality sign, and get all other terms on the other side.

To accomplish this, we use two rules:

Rule 1: **A term may be transposed from one side of the inequality to the other by changing its sign as it crosses the inequality sign.**

EXAMPLE 1    If $x + 5 < 7$,

then $x < 7 - 5$

Note sign change.

and $x < 2$.

EXAMPLE 2    If    $1 - x > -6$,

then $1 + 6 > x$

Note sign changes.

and     $7 > x$

or      $x < 7$.

Rule 2: **Both sides of an inequality may be multiplied or divided by the same nonzero number provided that:**
   **a.  if the number is positive, the direction of the inequality remains the same;**
   **b.  if the number is negative, the direction of the inequality is reversed.**

**EXAMPLE 3**    If    $6 < 15,$

then $\dfrac{6}{3} < \dfrac{15}{3}$

and $2 < 5.$

**EXAMPLE 4**    If    $\dfrac{1}{4} < 12,$

then $4\left(\dfrac{1}{4}\right) < 4(12)$

and    $1 < 48.$

**EXAMPLE 5**    If    $15 > 10,$

then $\dfrac{15}{-5} < \dfrac{10}{-5}$

and    $-3 < -2.$

> Note:    For the inequality to be true, the direction of the **inequality must be reversed** when the inequality is divided by the same negative number.

**EXAMPLE 6**    If    $-2x \le 8,$

then $\dfrac{-2x}{-2} \ge \dfrac{8}{-2}$

and    $x \ge -4.$

> Note again:    For the inequality to be true, the direction of the **inequality must be reversed** when the inequality is divided by the same negative number.

**EXAMPLE 7**    If    $4 > -2,$

then $(-1)4 < (-1)(-2)$

and    $-4 < 2.$

> Note:    In this example, the inequality is multiplied by the same negative number, and for the inequality to be true, the direction of the **inequality must be reversed**.

---

Now let's get on with the business of solving first-degree inequalities. The same four steps used to solve first-degree equations will again be used to solve inequalities. However, we must remember to change the direction of the inequality if we multiply or divide by the same negative number.

Recall the four steps:

1. **Simplify** by removing parentheses,
    clearing fractions,
    collecting like terms.
2. **Transpose.**
3. **Simplify.**
4. **Divide by the coefficient of the variable.**

Now, we must remember:

> **Reverse the direction** of an inequality symbol whenever an inequality is multiplied or divided by the same negative number.

Here are some examples that illustrate the use of these steps in solving first-degree inequalities. Read through each example and be sure you understand what has happened at each step.

## EXAMPLE 8

Solve:   $4x - 7 > 6x + 5$.

---

Solution:

$$4x - 7 > 6x + 5$$
$$4x - 6x > 5 + 7$$
$$-2x > 12$$
$$\frac{-2x}{-2} < \frac{12}{-2} \quad \text{Reverse direction.}$$
$$x < -6$$

Answer: $x < -6$ (or $-6 > x$).

To say that $x < -6$ is the answer to the inequality means that all numbers less than $-6$ satisfy the inequality. For example, if we substitute $x = -7$ into the inequality, we obtain

$$4(-7) - 7 > 6(-7) + 5$$
$$-28 - 7 > -42 + 5$$
$$-35 > -37$$

and, rewriting,           $-37 < -35$

which is true, so we say that $x = -7$ satisfies the inequality.

Graphically, $x < -6$ can be represented on the number line with an open circle at $-6$ and a heavy line to the left. The open circle indicates that $-6$ is not part of the answer. The heavy line indicates that all numbers to the left of $-6$ are part of the answer, including such numbers as $-6.1$ and $-6.25$.

---

## EXAMPLE 9

Solve:   $5(x + 3) \geq 31 + x$.

---

Solution: 

$$5(x + 3) \geq 31 + x$$
$$5x + 15 \geq 31 + x$$
$$5x - x \geq 31 - 15$$
$$4x \geq 16$$
$$x \geq 4$$

Answer:   $x \geq 4$ or, rewritten, $4 \leq x$

To say that $x \geq 4$ is the answer to the inequality means that all numbers greater than or equal to 4 will satisfy the inequality.

If the answer were graphed on a number line, there would be a closed circle at 4 and a heavy line to the right.

Now try to solve the inequalities in Problems 1 and 2. Remember to **reverse the direction of the inequality symbol** whenever the inequality is multiplied or divided by the same negative number.

## Problem 1

Solve:   $3x - x - 7 \geq (x - 2) + 4$.

Solution:

Answer:   $x \geq 9$ (or $9 \leq x$)

## Problem 2

Solve:   $x - 4 - 2(6 - x) > 2(3x - 5)$.

Solution:

Answer:   $x < -2$

Now, in contrast to our earlier examples, consider the following inequality.

## EXAMPLE 10

Solve:   $2x + 1 + x < 3(x + 2)$.

---

Solution:   $2x + 1 + x < 3(x + 2)$

$$1 + 3x < 3x + 6$$
$$3x - 3x < 6 - 1$$
$$0 < 5$$

Answer:   The solution is the entire set of real numbers.

---

The inequality in Example 10 is *always* true, regardless of the value of $x$, because 0 is always less than 5. Its solution is the entire set of real numbers. Remember, the real numbers are all the counting numbers, zero, whole numbers, positive and negative integers, fractions, rational numbers, and irrational numbers. In other words, $x$ can equal any number.

For example,              if $x = 0$, then $2(0) + 1 + (0) < 3[(0) + 2]$

and $1 < 6$.

Or:                 if $x = 5$, then $2(5) + 1 + (5) < 3[(5) + 2]$

and $10 + 1 + 5 < 3(7)$

$$16 < 21.$$

You should now be able to solve any first-degree inequality.

The same four basic steps—simplify, transpose, simplify, and divide—are used to solve first-degree inequalities. In addition, you must remember to reverse the direction of the inequality symbol whenever you multiply or divide an inequality by the same negative number.

Now try to solve the problems in the exercises.

## EXERCISES

Solve, and graph the answer for each problem on a number line:

1.   $10 + 2x \leq 12$

2.   $4 - (12 - 3x) \leq -5$

3.   $5x < 22 - (2x + 1)$

4.   $4x + (3x - 7) > 2x - (28 - 2x)$

5.   $5 - 3x \leq 23$

6.   $3x + 4(x - 2) \geq x - 5 + 3(2x - 1)$

7.   $3x - 2(5x + 2) > 1 - 5(x - 1) + x$

8.   $3x - 2(x - 5) < 3(x - 1) - 2x - 11$

9.   $3x + 4(x - 2) + 7 > x - 5 + 3(2x - 1)$

10.   $5x - 2(3x - 4) > 4[2x - 3(1 - 3x)]$     Be careful!

# UNIT 29

## Solving Quadratic Inequalities

In this unit we will continue our discussion of inequalities. When you have completed the unit, you will be able to solve quadratic inequalities in one variable.

### WRITING DOUBLE INEQUALITIES

A **double inequality** has two inequality symbols of the same sense combined in one statement.

For example, $2 < x < 5$ is a double inequality. In words, $2 < x < 5$ means that $x$ represents all numbers greater than 2 but less than 5. In other words, $x$ lies between 2 and 5 on the number line.

Graphically, $2 < x < 5$ is represented as follows:

Observe that $5 > x > 2$ has the same meaning as $2 < x < 5$. Both statements indicate that 5 is the larger number, 2 is the smaller number, and $x$ lies between them.

As stated in Unit 28, I prefer to write most inequalities using the $<$ symbol, because it is clearer to visualize the relationship of the numbers and the symbols on the number line. For example:

What does $-1 \leq x \leq 4$ mean? To say that $-1 \leq x \leq 4$ indicates that $x$ represents all numbers between $-1$ and $4$, including the numbers $-1$ and $4$. Graphically it would look like this:

# SECOND-DEGREE OR QUADRATIC INEQUALITIES

Recall from Unit 21:

---

Definition:   $ax^2 + bx + c = 0$, with $a$, $b$, and $c$ being real numbers, $a \neq 0$, is called a **second-degree equation** or **quadratic equation**.

---

A second-degree or quadratic inequality has the same characteristics as a second-degree equation, except that in place of the equal sign there is an inequality symbol. The standard form of the inequality has **all** terms on the left side of the inequality symbol with **only** 0 on the right.

Examples of second-degree or quadratic inequalities in standard form are:

$$x^2 - 3x + 5 < 0$$
$$7x^2 + 2x - 1 > 0$$
$$x^2 \geq 0$$
$$4x^2 + 7 \leq 0$$

whereas

$x^3 - x^2 > 0$ is a third-degree inequality

$x^2 + y < 0$ has two variables

$3x - 7 \leq x$ is a first-degree inequality

To solve a second-degree or quadratic inequality is to find the values of $x$ that satisfy the inequality. The basic technique developed here will rely heavily on our knowledge of graphing quadratic or second-degree equations.

Recall some of the basic facts (page 190) about the graph of a quadratic equation in standard form: $y = ax^2 + bx + c$.

The graph is a smooth, $\cup$ -shaped curve called a parabola.

If $a$ is positive, the curve opens up. If $a$ is negative, the curve opens down.

The $y$-intercept is at $c$.

There are at most two $x$-intercepts, which are found by solving $0 = ax^2 + bx + c$.

The procedure I will use to solve quadratic inequalities involves four major steps.

1. If necessary, first rewrite the inequality in standard form; that is, put all the terms on the left side with only 0 remaining on the right side of the inequality symbol.

2. Let $y =$ the algebraic expression on the left side of the inequality symbol, resulting in a quadratic **equation**.

3. Find the $x$-intercepts by solving $0 = ax^2 + bx + c$ and do a rough sketch of the quadratic equation showing only the intercept(s). There is no need to locate the vertex or any extra points.

4. By inspection of the graph, determine the answer to the inequality.

The examples that follow illustrate this procedure.

## EXAMPLE 1

Solve: $x^2 - 2x - 3 < 0$.

Solution:  Let $y = x^2 - 2x - 3$ and graph.

This is a quadratic equation; graph is a parabola.

Since $a = 1$ is positive, curve opens up.

The $y$-intercept is $c = -3$.

Find the $x$-intercepts, at most two, by solving:

$$0 = x^2 - 2x - 3$$
$$0 = (x + 1)(x - 3)$$
$$x = -1 \quad \text{or} \quad x = 3$$

The $x$-intercepts are $-1$ and $3$.

Do a rough sketch of the parabola, showing only the intercepts. There is no need to locate the vertex.

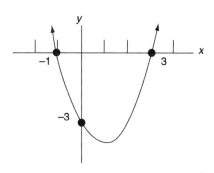

By inspection of the graph, answer this question:

For what values of $x$ is $x^2 - 2x - 3 < 0$?

Or, in other words:

For what values of $x$ is $y < 0$?

Or, in still other words:

For what values of $x$ is the curve below the $x$-axis?

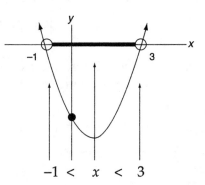

The curve is below the $x$-axis when $x$ is any number between $-1$ and $3$.

Answer:

$$-1 < \quad x \quad < \quad 3$$

To say that $-1 < x < 3$ is the answer to the inequality means that all numbers between $-1$ and $3$ satisfy the inequality. For example, if we substitute $x = 1$ into the inequality, we obtain

$$(1)^2 - 2(1) - 3 < 0$$
$$1 - 2 - 3 < 0$$
$$-4 < 0$$

which is true, so we say that $x = 1$ satisfies the inequality.

---

# EXAMPLE 2

Solve: $x^2 + 5x + 6 < 0$.

---

Solution: Let $y = x^2 + 5x + 6$ and graph.

This is a quadratic equation; graph is a parabola.

Since $a = 1$ is positive, curve opens up.

The $y$-intercept is $c = 6$.

Find the $x$-intercepts, at most two, by solving:

$$0 = x^2 + 5x + 6$$
$$0 = (x + 2)(x + 3)$$
$$x = -2 \quad \text{or} \quad x = -3$$

The $x$-intercepts are $-2$ and $-3$.

Do a rough sketch of the parabola, showing only the intercepts. There is no need to locate the vertex.

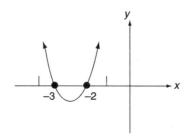

By inspection of the graph, answer this question:

For what values of $x$ is $x^2 + 5x + 6 < 0$?

Or:

For what values of $x$ is $y < 0$?

Or:

For what values of $x$ is the curve below the $x$-axis?

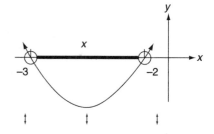

The curve is below the $x$-axis when $x$ is any number between $-2$ and $-3$.

Answer:                                    $-3 \; < \; x \; < \; -2$

-------------------------------------------------------------------

Try solving this problem.

## Problem 1

Solve:   $x^2 - 3x - 10 < 0$.

-------------------------------------------------------------------

Solution:

Answer:   $-2 < x < 5$

-------------------------------------------------------------------

I'll do another example, then you try two problems.

## EXAMPLE 3

Solve: $-x^2 + 4 < 0$.

Solution: Let $y = -x^2 + 4$ and graph.

Since $a = -1$ is negative, the parabola opens down.

The $y$-intercept is $c = 4$.

Find the $x$-intercepts, at most two, by solving:

$$0 = -x^2 + 4$$
$$x^2 = 4$$
$$x = -2 \quad \text{or} \quad x = 2$$

The $x$-intercepts are $-2$ and $2$.

Do a rough sketch of the parabola, showing only intercepts.

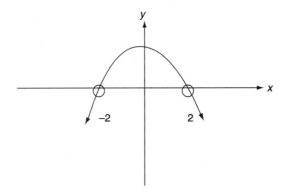

By inspection of the graph, answer this question:

For what values of $x$ is the curve below the $x$-axis?

In this example, the parabola is opening down, and the two ends of the curve are below the $x$-axis. Thus the curve is below the $x$-axis when $x$ is less than $-2$ or when $x$ is greater than 2.

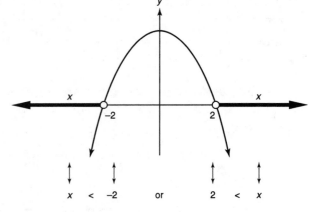

Answer: $x < -2$ or $2 < x$

Note: The answer cannot be written as a double inequality because $x$ does *not* lie between two numbers.

## Problem 2

Solve:   $-x^2 + 4x + 5 < 0$.

Solution:

Answer:   $x < -1$ or $5 < x$

## Problem 3

Solve:   $x^2 + 1 < 0$.

Solution:

Answer:   no solution

The entire graph is above the $x$-axis or, in other words, $x^2 + 1$ is always positive.

Thus far, all our examples have been inequalities with the $<$ symbol. When an inequality has a $>$ symbol, the procedure for solving it is similar except for the final question.

To solve an inequality, written in standard form, involving $>$, the final question becomes:

For what values of $x$ is $y > 0$?

Or, in other words:

For what values of $x$ is the curve **above** the $x$-axis?

## EXAMPLE 4

Solve: $-x^2 + x + 12 > 0$.

Solution: Let $y = -x^2 + x + 12$ and graph.

This is a quadratic equation; graph is a parabola.

Since $a = -1$ is negative, curve opens down.

The $y$-intercept is $c = 12$.

The $x$-intercepts:

$$0 = -x^2 + x + 12$$
$$0 = x^2 - x - 12$$
$$0 = (x - 4)(x + 3)$$
$$x = 4 \quad \text{or} \quad x = -3$$

The $x$-intercepts are $-3$ and $4$.

A rough sketch of the parabola is:

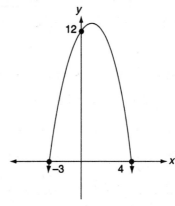

By inspection of the graph, answer this question:

For what values of $x$ is $-x^2 + x + 12 > 0$?

Or:

For what values of $x$ is $y > 0$?

Or:

For what values of $x$ is the curve **above** the $x$-axis?

Answer: $-3 < x < 4$

Here is an inequality with > for you to solve.

---

## Problem 4

Solve:   $-x^2 + 7x - 6 > 0$.

---

Solution:

Answer:   $1 < x < 6$

---

To generalize the situation:

> Given $ax^2 + bx + c < 0$ to solve, the final question to be answered is:
>
> For what values of $x$ is the curve **below** the $x$-axis?
>
> Given $ax^2 + bx + c > 0$ to solve, the final question to be answered is:
>
> For what values of $x$ is the curve **above** the $x$-axis?

Here is one more example in detail with explanation.

===

## EXAMPLE 5

Solve:   $3x^2 \geq -4x - 1$.

---

Solution:   Write in standard form: $3x^2 + 4x + 1 \geq 0$.

Let $y = 3x^2 + 4x + 1$ and graph.

This is a quadratic equation; graph is a parabola.

Since $a = 3$ is positive, curve opens up.

The $y$-intercept is $c = 1$.

The $x$-intercepts:

$$0 = 3x^2 + 4x + 1$$
$$0 = (3x + 1)(x + 1)$$
$$x = -\frac{1}{3} \quad \text{or} \quad x = -1$$

The $x$-intercepts are $-1$ and $-\frac{1}{3}$.

A rough sketch of the parabola is:

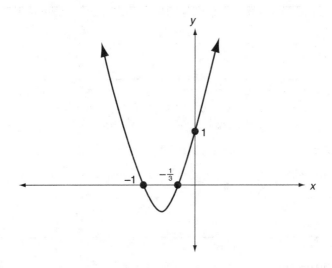

By inspection of the graph, answer this question:

For what values of $x$ is the curve **above** or **on** the $x$-axis?

Answer: $x \le -1$ or $-\dfrac{1}{3} \le x$

Note: The $x$-intercepts are part of the answer. Why?

Convince yourself, by substitution, that both $x = -1$ and $x = -\dfrac{1}{3}$ satisfy the inequality.

-----

You should now be able to identify and solve second-degree or quadratic inequalities in one variable. Remember our four-step procedure, which, simply stated, is:

1. Write in standard form.
2. Let $y =$ the algebraic expression on the left.
3. Draw a rough sketch showing only the intercept(s).
4. By inspection, determine the answer.

If the inequality involves <, the final question to be answered is:

For what values of $x$ is the curve **below** the $x$-axis?

If the inequality involves >, the final question to be answered is:

For what values of $x$ is the curve **above** the $x$-axis?

Also remember, if the inequality is combined with an = sign, the $x$-intercepts are part of the answer.

Before going on to the next unit, try solving the following exercises.

## EXERCISES

Solve for $x$ by sketching the graph of each quadratic equation. Or, optionally, view the graph of each quadratic equation on a graphing calculator and determine its $x$-intercept(s). Then by inspection of the graph, determine the solution to the inequality.

1.  $x^2 + 4x - 21 < 0$

2.  $x^2 + 2x - 8 < 0$

3.  $x^2 - x - 12 > 0$

4.  $-x^2 - 4x - 3 < 0$

5.  $-x^2 - x + 2 > 0$

6.  $-4x^2 \geq 0$

7.  $x^2 - 4x + 4 \leq 0$

8.  $4 + 2x^2 \leq 0$

9.  $9x^2 - 9x \leq 0$

10.  $25 - x^2 \geq 0$

11.  $x^2 - 4x + 13 > 0$

12.  $x^2 + 6x + 9 > 0$

13.  $-x^2 + 2x + 15 \leq 0$

14.  $2x^2 + 4x - 1 < 0$

15.  $2x^2 + 1 \leq 0$

16.  $-x^2 + 10x \leq 0$

17.  $2x^2 + 3x - 5 \leq 0$

18.  $3x^2 - 3x + 2 > 0$

19.  $-6x^2 - x + 2 \geq 0$

20.  $2x^2 - 12x + 3 \leq 0$

# UNIT 30

## Logarithms

The purpose of this unit is to provide you with a brief overview of logarithms. First the definition and notation used with logarithms will be introduced. Then you will learn about logarithmic properties and the ways these properties are used to simplify expressions involving logarithms.

The equation $\log_b N = x$ is read "the logarithm of $N$ to the base $b$ is $x$."

---

Definition:  $x$ is called the **logarithm of $N$ to the base $b$** if $b^x = N$, where $N$ and $b$ are both positive numbers, $b \neq 1$.

---

In other words:

$$\log_b N = x \quad \text{if and only if} \quad b^x = N.$$

Examples of logarithms are:

$$\log_3 9 = 2 \quad \text{because} \quad 3^2 = 9$$
$$\log_2 8 = 3 \quad \text{because} \quad 2^3 = 8$$
$$\log_7 7 = 1 \quad \text{because} \quad 7^1 = 7$$
$$\log_5 25 = 2 \quad \text{because} \quad 5^2 = 25$$
$$\log_{13} 1 = 0 \quad \text{because} \quad 13^0 = 1$$
$$\log_2 \frac{1}{2} = -1 \quad \text{because} \quad 2^{-1} = \frac{1}{2}$$

# LOGARITHMIC AND EXPONENTIAL FORMS

Notice that **the logarithm of a positive number $N$ is the exponent to which the base must be raised to produce the number $N$.**

$$\log_b N = x \text{ is called the logarithmic form,}$$
$$b^x = N \text{ is called the exponential form,}$$

and the two statements are equivalent.

Try to develop your ability to go from one form to the other.

## EXAMPLE 1

Express $\log_8 16 = \dfrac{4}{3}$ in exponential form.

- - - - - - - - - - - - - - - - - - - - - - - - - - - - - - - - - - - - - - - -

Solution:   If   $\log_8 16 = \dfrac{4}{3}$,

then   $8^{4/3} = 16$.

- - - - - - - - - - - - - - - - - - - - - - - - - - - - - - - - - - - - - - - -

Cover the answers, and then try to express each of the following in exponential form.

**Problem 1**   $\log_3 243 = 5$

**Problem 2**   $\log_2 16 = 4$

**Problem 3**   $\log_{11} 121 = 2$

**Problem 4**   $\log_u w = y$

Answers:   1. $3^5 = 243$   2. $2^4 = 16$   3. $11^2 = 121$   4. $u^y = w$

Students usually find it more difficult to go from exponential form to logarithmic form. I start by writing the base first, then the exponent. Remember that the logarithm of a number is an **exponent.**

## EXAMPLE 2

Express $5^3 = 125$ in logarithmic form.

- - - - - - - - - - - - - - - - - - - - - - - - - - - - - - - - - - - - - - - -

Solution:   If           $5^3 = 125$,

then   $\log_5 125 = 3$.

- - - - - - - - - - - - - - - - - - - - - - - - - - - - - - - - - - - - - - - -

Cover the answers, and then try to express each of the following in logarithmic form.

---

## Problem 5     $4^2 = 16$

## Problem 6     $12^0 = 1$

## Problem 7     $4^{-5/2} = \dfrac{1}{32}$

## Problem 8     $A^2 = c$

Answers:    5. $\log_4 16 = 2$     6. $\log_{12} 1 = 0$     7. $\log_4 \dfrac{1}{32} = -\dfrac{5}{2}$    8. $\log_A c = 2$

---

Actually any positive number can be used for the base. Logarithms to the base 10 are called **common logs**, and the 10 is omitted from the logarithmic notation. In other words, if no base is written using logarithmic notation, the base is understood to be 10.

With common logs "the logarithm of $N$" is often shortened and read as "log $N$."

## EXAMPLE 3

Express $\log A = c$ in exponential form.

---

Solution:    $\log A = c$ denotes a common logarithm with base 10 because no base number is indicated.

           Therefore,   if   $\log A = c$,

                 then    $10^c = A$.

---

It's your turn now.

---

## Problem 9

Express $\log 100 = 2$ in exponential form.

---

Solution:

Answer:    $10^2 = 100$

---

# SOLVING LOGARITHMIC EQUATIONS

Now let's try solving some logarithmic equations of the form $x = \log_b N$, where we are to find the unknown number.

If you are more comfortable with the exponential form at this stage, I suggest changing from the logarithmic form to the exponential form to solve the equations. Eventually, though, you will need to solve the equations in their original form.

## EXAMPLE 4

Solve:   $\log_7 7 = x$.

Solution:   If      $\log_7 7 = x$,

then      $7^x = 7$

and      $7^x = 7^1$   since $7 = 7^1$.

Answer:   $x = 1$

Here's an alternative method, which I think is easier:

In words, $\log_7 7 = x$ means "7 raised to what power $x$ equals 7?"

The answer is 1.

## EXAMPLE 5

Solve:   $\log_2 32 = x$.

Solution:   If   $\log_2 32 = x$,

then      $2^x = 32$

and      $2^x = 2^5$   since $32 = 2^5$.

Answer:   $x = 5$

Alternative method:

In words, $\log_2 32 = x$ means "2 raised to what power $x$ equals 32?"

The answer is 5.

Try Problems 10–12 on your own.

## Problem 10

Solve:   $\log_8 64 = x$.

Solution:

Alternative method:

In words, $\log_8 64 = x$ means "8 raised to what power $x$ equals 64?"

Answer:   $x = 2$

## Problem 11

Solve:   $\log 1000 = x$.

Solution:

Answer:   $x = 3$

## Problem 12

Solve:   $\log_7 1 = x$.

Solution:

Answer:   $x = 0$

In the preceding examples and problems, the base and number were given and we had to find the logarithm. The following examples illustrate the technique to be used when the log is given and we wish to solve for either the base or the number.

## EXAMPLE 6

Solve:  $\log_3 x = 4$.

Solution:  If  $\log_3 x = 4$,

then  $3^4 = x$

and  $81 = x$   since $3^4 = 81$.

Answer:  $x = 81$

## EXAMPLE 7

Solve:  $\log_x 9 = \dfrac{1}{2}$.

Solution:  If  $\log_x 9 = \dfrac{1}{2}$,

then  $x^{1/2} = 9$

or  $\sqrt{x} = 9$.

Answer:  $x = 81$

Since Example 7 may be a bit confusing, let's do one more before you try Problems 13–17.

## EXAMPLE 8

Solve:  $\log_x 64 = 6$.

Solution:  If  $\log_x 64 = 6$,

then  $x^6 = 64$

and  $x^6 = 2^6$   since $64 = 2^6$.

Answer:  $x = 2$

## Problem 13

Solve: $\log_5 x = 1$.

Solution:

Answer: $x = 5$

## Problem 14

Solve: $\log x = 4$.

Solution:

Answer: $x = 10,000$

## Problem 15

Solve: $\log_3 x = -2$.

Solution:

Answer: $x = \dfrac{1}{9}$

## Problem 16

Solve:   $\log_x 100 = 2$.

Solution:

Answer:   $x = 10$

## Problem 17

Solve:   $\log_x 8 = 3$.

Solution:

Answer:   $x = 2$

By now you should be asking such questions as:

These problems were okay, but what if the answer is not obvious?
What if I have something other than an integer value for the log?
What if the number is not an integral power of the base?

The next example is intended to answer most of your questions.

## EXAMPLE 9

Solve:   $\log 5 = x$.

Solution:   If      $\log 5 = x$,
            then     $10^x = 5$.

Problems like this, where $x$ is not an integral value, cannot be solved by elementary algebra. It is possible to find a decimal approximation for the common log of any positive number through the use of the log key on most calculators. However, problems of this type are beyond the scope of this unit and therefore will not be considered.

# SIMPLIFYING EXPRESSIONS WITH LOGARITHMS

To accomplish simplification, we have **four** basic properties of logarithms.

These properties are used to shorten computations or to simplify complicated expressions involving products, quotients, powers, and roots.

---

**Properties of Logarithms**

Product property: $\log_b AC = \log_b A + \log_b C$

Quotient property: $\log_b \dfrac{A}{C} = \log_b A - \log_b C$

Power property: $\log_b A^k = k \log_b A$

Root property: $\log_b \sqrt[k]{A} = \dfrac{1}{k} \log A$

---

The following examples and problems illustrate how these four properties can be used to shorten computations. Remember that the logarithm of a positive number $N$ is the exponent to which the base must be raised to produce the number $N$.

---

Product property: $\log_b AC = \log_b A + \log_b C$

---

Verbally, the product property states that the logarithm of a product of two numbers is equal to the sum of the logarithms of the numbers.

---

## EXAMPLE 10

Find: $\log_3(81 \cdot 9)$.

- - - - - - - - - - - - - - - - - - - - - - - - - - - - - - - - - - - - - - -

Solution: $\log_3(81 \cdot 9) = \log_3 81 + \log_3 9$

$\qquad\qquad\quad = 4 + 2$

$\qquad\qquad\quad = 6$

Answer: $\log_3(81 \cdot 9) = 6$

In case you are not convinced:

If $\quad \log_3(81 \cdot 9) \overset{?}{=} 6,$

then $\qquad\qquad 3^6 \overset{?}{=} (81 \cdot 9)$

and $\qquad\qquad 729 = 729.$

- - - - - - - - - - - - - - - - - - - - - - - - - - - - - - - - - - - - - - -

## Problem 18

Find: $\log_2(8 \cdot 4)$.

Solution: Use the product property.

Answer: 5

## Problem 19

Find: $\log_2(64 \cdot 32)$.

Solution: Use the product property.

Answer: 11

$$\text{Quotient property:} \quad \log_b \frac{A}{C} = \log_b A - \log_b C$$

In words, the quotient property states that the logarithm of a quotient of two numbers $A$ and $C$ is equal to the difference of the logarithms of the numbers $A$ and $C$.

## EXAMPLE 11

Find: $\log_3 \frac{1}{27}$.

Solution: $\log_3 \frac{1}{27} = \log_3 1 - \log_3 27$

$$= 0 - 3$$
$$= -3$$

Answer: $\log_3 \frac{1}{27} = -3$

If you are not convinced, change from the logarithmic form to the **exponential** form and verify that both sides represent the same number.

## EXAMPLE 12

Question:   If $\log 3 = 0.477$ and $\log 2 = 0.301$, what does $\log 1.5$ equal?

Solution:   $\log 1.5 = \log \dfrac{3}{2}$

$\qquad\qquad = \log 3 - \log 2$

$\qquad\qquad = 0.477 - 0.301$

$\qquad\qquad = 0.176$

Answer:   $\log 1.5 = 0.176$

If you are not convinced, use the log key on your calculator to verify that the log of 1.5 is 0.176.

## Problem 20

Find:   $\log_2 \dfrac{4}{64}$.

Solution:   Use the quotient property.

Answer:   $-4$

## Prroblem 21

Find:   $\log_2 \dfrac{1}{4}$.

Solution:   Use the quotient property.

Answer:   $-2$

$$\text{Power property:} \quad \log_b A^k = k \log_b A$$

In words, the power property states that the logarithm of a power of a number $A$ is equal to the power times the logarithm of the number $A$.

## EXAMPLE 13

Find:   $\log_3(81)^5$.

Solution:   $\log_3(81)^5 = 5\log_3 81$
$$= 5(4)$$
$$= 20$$

$$\text{Root property:} \quad \log_b \sqrt[k]{A} = \frac{1}{k}\log_b A$$

The root property follows directly from the power property. In words, the root property states that to take the logarithm of a root of a number $A$, rewrite the root of the number $A$ using a fractional exponent and then apply the power property.

The following example is an illustration of the use of this property.

## EXAMPLE 14

Find:   $\log_3 \sqrt{27}$.

Solution:   $\log_3 \sqrt{27} = \log_3(27)^{1/2}$
$$= \frac{1}{2}\log_3 27$$
$$= \frac{1}{2}(3)$$
$$= 1.5$$

---

## Problem 22

Find:  $\log_2(32)^3$.

---

Solution:  Use the power property.

Answer:  15

---

---

## Problem 23

Find:  $\log_2 \sqrt{32}$.

---

Solution:  Use the root property.

Answer:  2.5

---

As previously stated, the four basic properties of logarithms are used also to simplify complicated expressions involving products, quotients, powers, and roots.

Let's try some dandy examples and problems.

---

## EXAMPLE 15

Simplify:  $\log\left(\dfrac{xy}{z}\right)$.

---

Solution:  $\log\left(\dfrac{xy}{z}\right) = \log(xy) - \log z$     quotient property

$= \log x + \log y - \log z$     product property

---

## EXAMPLE 16

Simplify:   $\log\left(\dfrac{x^3 y}{z}\right)$.

Solution:   $\log\left(\dfrac{x^3 y}{z}\right) = \log(x^3 y) - \log z$        quotient property

$$= \log x^3 + \log y - \log z$$        product property
$$= 3\log x + \log y - \log z$$        power property

## EXAMPLE 17

Simplify:   $\log\sqrt{xy}$ with $x > 0$ and $y > 0$.

Solution:   $\log\sqrt{xy} = \log(xy)^{1/2}$

$$= \frac{1}{2}\log(xy)$$        root property

$$= \frac{1}{2}(\log x + \log y)$$        product property

$$= \frac{1}{2}\log x + \frac{1}{2}\log y$$

The last two problems are for you.

## Problem 24

Simplify:   $\log(x^2 y)$.

Solution:

Answer:   $2\log x + \log y$

## Problem 25

Simplify:   $\log\sqrt{\dfrac{x}{y}}$ with $x > 0$ and $y > 0$.

Solution:

Answer:   $\dfrac{1}{2}(\log x - \log y)$

You should now have a basic understanding of logarithms. Remember that the logarithm of a positive number $N$ is the exponent to which the base must be raised to produce the number $N$.

Also, you should be familiar with both logarithmic notation and exponential notation and be able to go from one form to the other.

Finally, you should be able to simplify logarithmic expressions involving products, quotients, and powers using the basic properties of logarithms.

Before beginning the next unit you should work the following exercises.

## EXERCISES

Solve for $x$:

1.  $\log_3 81 = x$

2.  $\log_5 125 = x$

3.  $\log_7\left(\dfrac{1}{7}\right) = x$

4.  $\log 1 = x$

5.  $\log_3 x = 2$

6.  $\log_7 x = 0$

7.  $\log_9 x = \dfrac{1}{2}$

8.  $\log_x 27 = 3$

9.  $\log_x 49 = 2$

10. $\log_x 121 = 2$

Express each of the given logarithms in terms of logarithms of $x$, $y$, and $z$, where the variables are all positive.

11. $\log x^5$

12. $\log 2xy^3$

13. $\log \dfrac{x^2}{y}$

14. $\log \dfrac{x}{yz}$   Be careful!

15. $\log \sqrt{x^3 y}$

# UNIT 31

## Right Triangles

The purpose of this, the last unit, is to provide you with a working knowledge of two frequently encountered triangles—the 30°-60°-90° triangle and the isosceles right triangle. When you have finished this unit, you will be able to recognize each of these triangles and, given the length of the one side, to find the lengths of the other two sides.

You probably recall that triangles are labeled using three capital letters, one at each vertex. The triangle shown below is referred to as triangle $ABC$ or, simply, $\triangle ABC$. The sides are labeled using lowercase letters as follows:

The side opposite angle $A$ is $a$.
The side opposite angle $B$ is $b$.
The side opposite angle $C$ is $c$.

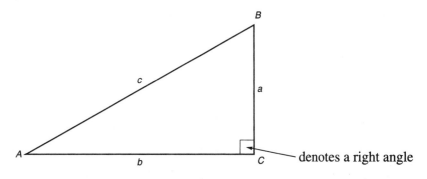

denotes a right angle

In any triangle the sum of the angles is equal to 180°. An angle of 90° is called a right angle. A **right triangle** is a triangle with a 90° angle. Thus triangle $ABC$ is a right triangle.

In a right triangle the side opposite the right angle is called the **hypotenuse**. In triangle $ABC$, angle $C$ is the right angle and $c$ is the hypotenuse. The hypotenuse is longer than either of the other two sides.

One of the most useful theorems with regard to right triangles is the Pythagorean theorem. More than likely, you can recall some version of it yourself.

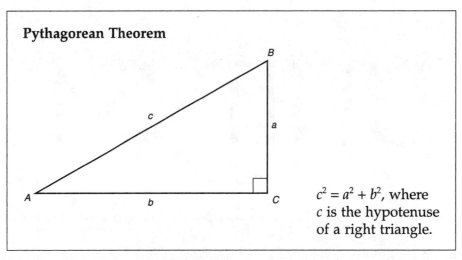

**Pythagorean Theorem**

$c^2 = a^2 + b^2$, where $c$ is the hypotenuse of a right triangle.

Or, in words, given a right triangle, the square of the length of the hypotenuse is equal to the sum of the squared lengths of the sides.

Be careful. The Pythagorean theorem is applicable only to right triangles.

## EXAMPLE 1

Question:   Given a right triangle with sides of 5 and 12, what is the length of the hypotenuse?

Solution:   Begin by drawing a picture.

Label what is given in the problem.

Let $a = 5$ and $b = 12$.

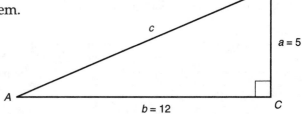

Because we are dealing with a right triangle, the Pythagorean theorem is applicable:

$$c^2 = a^2 + b^2$$
$$= (5)^2 + (12)^2$$
$$= 25 + 144$$
$$= 169$$
$$c = \sqrt{169}$$
$$= 13$$

Answer:   The length of the hypotenuse is 13.

Note:   Only the positive square root of 169 is the answer, because $c$ represents the length of a side of a triangle. A negative number would be meaningless for an answer in this situation.

## EXAMPLE 2

Question:   The hypotenuse of a right triangle is 3, and a side adjacent to it is 2. What is the length of the third side?

Solution:   Begin by drawing a picture.

Label what is given in the problem.

Let $c = 3$ and $a = 2$.

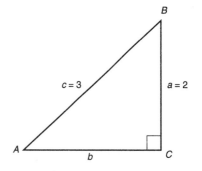

By the Pythagorean theorem:

$$c^2 = a^2 + b^2$$
$$(3)^2 = (2)^2 + b^2$$
$$9 = 4 + b^2$$
$$5 = b^2$$
$$\sqrt{5} = b$$

Answer:   The length of the third side of the triangle is $\sqrt{5}$ or, if an approximation is sufficient, 2.236.

I am sure you are ready to try a few problems now.

## Problem 1

Question:   One side of a right triangle is 7, and the other side is 5. What is the length of the hypotenuse?

Solution:

Answer:   $\sqrt{74}$

## Problem 2

Question:  The hypotenuse of a right triangle is 5, and an adjacent side is 3. What is the length of the third side?

Solution:

Answer: 4

Notice that, when you are given the lengths of **any two sides of a right triangle**, the length of the third side can be found by using the Pythagorean theorem.

## Problem 3

Question:  The longest side of a right triangle is 11, and the shortest side is 3. What is the length of the other side?

Solution:

Answer: $\sqrt{112}$

# 30°-60°-90° TRIANGLES

One of the most frequently encountered right triangles is the **30°-60°-90° triangle**, so named because

one angle is 30°,
another angle is 60°, and
the third angle is 90°.

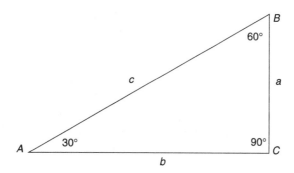

In a 30°-60°-90° triangle the sides have the following relationships:

---

The length of the **side opposite the 30° angle** is equal to half the length of the hypotenuse, or, stated as a formula,

$$a = \frac{c}{2}$$

The length of the **side opposite the 60° angle** is equal to half the length of the hypotenuse times $\sqrt{3}$, or, stated as a formula,

$$b = \frac{c}{2}\sqrt{3}$$

---

## EXAMPLE 3

Question:  If the hypotenuse of a 30°-60°-90° triangle is 6, what are the lengths of the other sides?

---

Solution:  Begin by drawing a picture.

Label what is given in the problem.

Given $c = 6$.

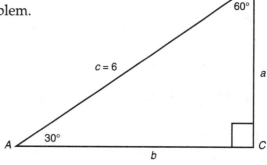

The length of the side opposite 30° is half the hypotenuse.

$$a = \frac{c}{2}$$
$$= \frac{6}{2}$$
$$= 3$$

The length of the side opposite 60° is half the hypotenuse times $\sqrt{3}$.

$$b = \frac{c}{2}\sqrt{3}$$
$$= \frac{6}{2}\sqrt{3}$$
$$= 3\sqrt{3}$$

Answer:   $a = 3$ and $b = 3\sqrt{3}$, located as shown in the drawing

------

Try Problems 4–6.

------

## Problem 4

Question:   Given the drawing at the right, what are the lengths of $a$ and $b$?

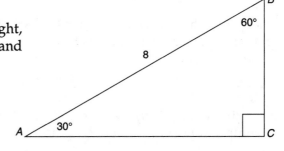

------

Solution:

Answer:   $a = 4$ and $b = 4\sqrt{3}$

------

Now suppose we change the conditions around a bit.

## Problem 5

Question:  If the length of the side opposite a 30° angle in a right triangle is 10, what is the length of the hypotenuse?

Solution:

Answer:  $c = 20$

## Problem 6

Question:  If the side opposite a 30° angle in a right triangle is 17, what is the length of the side opposite the 60° angle?

Solution:

Answer:  $17\sqrt{3}$

In the next example I will do the more difficult situation where the length of the side opposite the 60° angle is given.

# EXAMPLE 4

Question: If the side opposite the 60° angle in a right triangle is 8, what are the lengths of the other two sides?

_____

Solution: We are given $b = 8$.

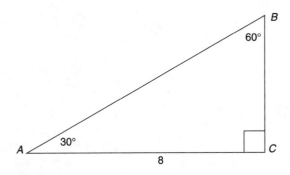

And we know that the side opposite the 60° angle is equal to half the hypotenuse times $\sqrt{3}$, or

$$b = \frac{c}{2}\sqrt{3}$$

So, by substitution,

$$8 = \frac{c}{2}\sqrt{3}$$

$$16 = c\sqrt{3}$$

$$\frac{16}{\sqrt{3}} = c$$

Or, if you prefer the denominator rationalized,

$$c = \frac{16\sqrt{3}}{\sqrt{3} \cdot \sqrt{3}}$$

$$= \frac{16\sqrt{3}}{3}$$

Once the hypotenuse is found, the side opposite the 30° angle is simply half of it.

$$a = \frac{8}{\sqrt{3}} \quad \text{or, if you prefer,} \quad \frac{8\sqrt{3}}{3}$$

Answer: The hypotenuse is $16/\sqrt{3}$, and the other side is $8/\sqrt{3}$.

_____

Here's a similar one for you.

## Problem 7

Question:   If the side opposite the 60° angle of a right triangle is 2, what are the lengths of the other two sides?

Solution:

Answer:   The hypotenuse is $4/\sqrt{3}$ or $4\sqrt{3}/3$,
and the short side is $2/\sqrt{3}$ or $2\sqrt{3}/3$.

# ISOSCELES RIGHT TRIANGLES

A triangle with two equal sides is called an **isosceles triangle**.

Another frequently encountered right triangle is the **isosceles right triangle**, so named because

two sides are equal, and
the angle between them is 90°.

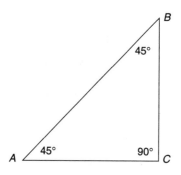

In an isosceles right triangle, if $c$ is the hypotenuse, $a = b$ and angle $A$ = angle $B$ = 45°. The two equal angles of an isosceles triangle are often referred to as the **base angles**.

As in the 30°-60°-90° triangle, the sides of an **isosceles right triangle** are related.

In an isosceles right triangle the sides have the following relationships:

---

The **sides opposite the 45° angles** are equal. Each side is equal to half the hypotenuse times $\sqrt{2}$, or, stated as a formula,

$$b = \frac{c}{2}\sqrt{2}, \quad \text{with } a = b$$

---

## EXAMPLE 5

Question: If an isosceles right triangle has a hypotenuse of 20, what are the lengths of the other sides?

---

Solution: As before, begin by drawing a picture.

Label what is given in the problem.

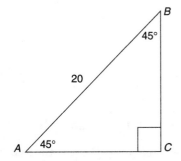

Because this is an isosceles right triangle, angle $A$ and angle $B$ are both 45° and $a = b$. Also, $c = 20$.

In an isosceles right triangle, the side opposite the 45° angle is half the hypotenuse times $\sqrt{2}$; therefore

$$b = \frac{c}{2}\sqrt{2}$$

and, by substitution,

$$b = \frac{20}{2}\sqrt{2}$$
$$= 10\sqrt{2}$$

Answer: The two equal sides are $10\sqrt{2}$.

---

This time you try the harder problem.

## Problem 8

Question:   If the side opposite a 45° angle in a right triangle is 5, what is the length of the hypotenuse?

Solution:

Answer:   $c = \dfrac{10}{\sqrt{2}}$ or $5\sqrt{2}$

Did you observe that, when working with either the 30°-60°-90° triangle or the isosceles right triangle, if you are given only the length of one side it is possible to find the lengths of the other two? In contrast, the Pythagorean theorem requires that two sides be known in order to find the third.

The preceding examples have been fairly easy, as you probably guessed, and the problems that you are likely to encounter rarely appear in such a straightforward manner. Instead, the information is contained in a word problem describing some real-life situation. However, if you remember to draw a diagram and label what is given in the problem, solving it should not be any more difficult than solving the problems you have done already.

A few word problems follow to illustrate what I mean.

## EXAMPLE 6

A ladder leans against the side of the building with its foot 7 ft from the building. If the ladder makes an angle of 60° with the ground, how long is the ladder?

Solution:   Begin by drawing a diagram.

Label what is given in the problem.

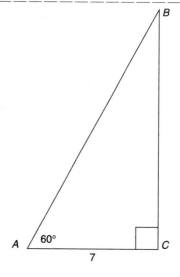

Angle $A = 60°$ and $b = 7$.

Therefore angle $B$ must be 30°. To find the length of the ladder, we apply our knowledge that the side opposite the 30° angle is half the hypotenuse. Since $b = 7$, $c$ must be 14.

Answer:   The ladder is 14 ft long.

---

## EXAMPLE 7

When the sun is 45° above the horizon, how long is the shadow cast by a tree 40 ft high?

---

Solution:   We are given angle $A = 45°$ and $a = 40$.

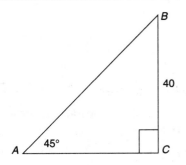

Since this is a right triangle, angle $B = 45°$ and we have an isosceles right triangle. Therefore $b$ must be 40.

Answer:   The shadow is 40 ft long.

---

## EXAMPLE 8

From the top of a tree, 25 ft above the ground, the angle of depression to an observer on the ground is 30°. How far is the tree from the observer?

---

Solution:   I will draw the diagram and label what is given.

30° = angle of depression

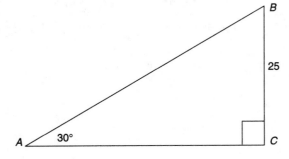

You can finish the problem from here.

Answer:   The tree is $25\sqrt{3}$ ft away.

_____

## EXAMPLE 9

Find the height of a street lamp if the angle of elevation of its top is 45° to an observer on the ground at a distance of 25 ft from its base.

_____

Solution:   Again I will draw the diagram and label what is given; then you finish the problem.

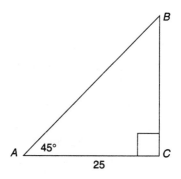

Answer:   The street lamp is 25 ft high.

_____

Try Problem 9 on your own without any clues.

## Problem 9

A man drives 1000 ft along a road that is inclined 30° to the horizontal. How high above his starting point is he?

_ _ _ _ _ _ _ _ _ _ _ _ _ _ _ _ _ _ _ _ _ _ _ _ _ _ _

Solution:

Answer:  He is 500 ft above his starting point.

_ _ _ _ _ _ _ _ _ _ _ _ _ _ _ _ _ _ _ _ _ _ _ _ _ _ _

You probably did fine on that problem, but let's do one more.

## Problem 10

A flagpole broken by the wind forms a right triangle with the ground. If the broken part makes an angle of 45° with the ground, and if the top of the flagpole is now 10 ft from its base, how tall was the flagpole?

_ _ _ _ _ _ _ _ _ _ _ _ _ _ _ _ _ _ _ _ _ _ _ _ _ _ _

Solution:

Answer:  The flagpole was $10 + 10\sqrt{2}$ ft tall.

_ _ _ _ _ _ _ _ _ _ _ _ _ _ _ _ _ _ _ _ _ _ _ _ _ _ _

You should now be able to define and label a right triangle. Remember that, when you are given the lengths of any two sides of a right triangle, the length of the third side can be found by using the Pythagorean theorem, $c^2 = a^2 + b^2$, where $c$ is the hypotenuse.

In addition you should be able to recognize two well-known right triangles.

1. **The 30°-60°-90° triangle** with properties:
   a.  the side opposite the 30° angle is equal to one-half of the hypotenuse;
   b.  the side opposite the 60° angle is equal to one-half of the hypotenuse times $\sqrt{3}$.

2. **The isosceles right triangle** with properties:
   a.  the sides opposite the 45° angles are equal;
   b.  the sides opposite the 45° angles are equal to one-half of the hypotenuse times $\sqrt{2}$.

Remember that, when working with either a 30°-60°-90° triangle or an isosceles right triangle, you need to be given only the length of one side in order to find the lengths of the other two sides. Also remember that the suggested procedure when solving such problems is to begin by drawing a diagram; then label what is given in the problem.

Congratulations; you have completed the book.

I can no longer say, "before continuing to the next unit," but do the following exercises anyway. There are only ten. Complete solutions are given at the end of the book for all ten exercises.

## EXERCISES

Solve:

1.  The hypotenuse of a right triangle is 7, and a side adjacent to it is 3. Find the length of the third side.

2.  Two sides of a triangle are 1 and 2. Find the length of the hypotenuse.

3.  If the hypotenuse of a 30°-60°-90° triangle is 10, find the lengths of the other sides.

4.  If the side opposite the 60° angle in a right triangle is 5, what are the lengths of the other two sides?

5.  If the side opposite the 30° angle in a right triangle is 1, what are the lengths of the other two sides?

6.  If the side opposite the 45° angle in a right triangle is 6, what are the lengths of the other two sides?

7.  If the hypotenuse of an isosceles right triangle is 10, what are the lengths of the other two sides?

8.  A ladder 20 ft long is resting against a house. If it makes an angle of 45° with the ground, how high up does it reach?

9.  An observer notes that the angle of elevation of the top of a neighboring building is 60°. If the distance from the observer to the neighboring building is 40 ft, how tall is the building?

10. When the sun is 30° above the horizon, an observer's shadow is 9 ft. How tall is the observer?

# Answers to Exercises

Note: For all units except Unit 1, detailed solutions are given for even-numbered problems. For Units 14, and 17, 30, and 31, detailed solutions are provided for all exercises.

## UNIT 1

1. $-7$
2. $4$
3. $-9$
4. $-1$
5. $-48$
6. $-18$
7. $44$
8. $-4$
9. $5$
10. $1$
11. $-1$
12. $-3$
13. $9$
14. $-3$
15. $-6$
16. $-10$
17. $-24$

18. $32$
19. $0$
20. $1$
21. $8$
22. $12 \div 3 \cdot 2 - 1 = 4 \cdot 2 - 1$
$$= 8 - 1$$
$$= 7$$
23. $1$
24. $(-10) \cdot 12 - 6 \cdot 3 = -120 - 18$
$$= -138$$
25. $6$
26. $-|-3| + 7 = -3 + 7$
$$= 4$$

## UNIT 2

1. $-7x - 14y + 21$

2. $3x + 4x + (x + 2)$

   $3x + 4x + x + 2$

   $8x + 2$

3. $9x + 2y + 3$

4. $-(-2x + 1) + 1$

   $2x - 1 + 1$

   $2x$

5. $9x + 3y + 1$

6. $(3x + 5xy + 2y) + (4 - 3xy - 2x)$

   $3x + 5xy + 2y + 4 - 3xy - 2x$

   $x + 2xy + 2y + 4$

7. $7y - 8a - 13$

8. $3(2a - b) - 4(b - 2a)$

   $6a - 3b - 4b + 8a$

   $14a - 7b$

9.  $14x + y - 17$

10. $5 - 2(x + 2[3 + x])$

$5 - 2(x + 6 + 2x)$

$5 - 2x - 12 - 4x$

$-6x - 7$

11. $-3x + 2$

12. $4 - 9(2x - 3) + 7(x - 1)$

$4 - 18x + 27 + 7x - 7$

$24 - 11x$

13. $8x + 4$

14. $4x - [5 - 3(2x - 6)]$

$4x - [5 - 6x + 18]$

$4x - 5 + 6x - 18$

$10x - 23$

---

# UNIT 3

1.  $x = 2$

2.  $2(x + 1) - 3(4x - 2) = 6x$

$2x + 2 - 12x + 6 = 6x$

$-10x + 8 = 6x$

$8 = 6x + 10x$

$8 = 16x$

$\dfrac{8}{16} = \dfrac{16x}{16}$

$\dfrac{1}{2} = x$

3.  $x = 23$

4.  $20 - \dfrac{3x}{5} = x - 12$

$5\left( 20 - \dfrac{3x}{5} = x - 12 \right)$

$100 - \dfrac{5 \cdot 3x}{5} = 5x - 60$

$100 - 3x = 5x - 60$

$-3x - 5x = -60 - 100$

$-8x = -160$

$\dfrac{-8x}{-8} = \dfrac{-160}{-8}$

$x = 20$

5.  $x = 2$

6.  $2(x + 2) = 5 + \dfrac{x + 1}{3}$

$2x + 4 = 5 + \dfrac{x + 1}{3}$

$3\left( 2x + 4 = 5 + \dfrac{x + 1}{3} \right)$

$6x + 12 = 15 + \dfrac{3(x + 1)}{3}$

$6x + 12 = 15 + x + 1$

$6x - x = 16 - 12$

$5x = 4$

$\dfrac{5x}{5} = \dfrac{4}{5}$

$x = \dfrac{4}{5}$

7.  $x = 6$

8.  $3 - \dfrac{5(x-1)}{2} = x$

$3 - \dfrac{5x-5}{2} = x$

Notice that the minus sign in front of the fraction remains, as I removed the parentheses by multiplying by a positive 5.

$2\left(3 - \dfrac{5x-5}{2} = x\right)$

$6 - \dfrac{\cancel{2}(5x-5)}{\cancel{2}} = 2x$

$6 - 5x + 5 = 2x$

$11 - 5x = 2x$

$11 = 2x + 5x$

$11 = 7x$

$\dfrac{11}{7} = \dfrac{\cancel{7}x}{\cancel{7}}$

$\dfrac{11}{7} = x$

$x = \dfrac{11}{7}$

9.  $x = -\dfrac{1}{3}$

10.  $1 - \dfrac{x}{2} = 5$

$2\left(1 - \dfrac{x}{2} = 5\right)$

$2 - x = 10$

$-x = 10 - 2$

$-x = 8$

$x = -8$

11.  $w = 0$

12.  $3(-2x + 1) = -6x - 7$

$-6x - 3 = -6x - 7$

$-6x + 6x = -7 + 3$

$0 = -4$

But 0 does not equal –4; therefore the problem has no solution.

13.  An identity, and the solution is all real numbers.

14.  $3(z+5) - 2z = \dfrac{z-1}{2} + 17$

$3z + 15 - 2z = \dfrac{z-1}{2} + 17$

$z + 15 = \dfrac{z-1}{2} + 17$

$2\left(z + 15 = \dfrac{z-1}{2} + 17\right)$

$2z + 30 = \dfrac{\cancel{2}(z-1)}{\cancel{2}} + 34$

$2z + 30 = z - 1 + 34$

$2z - z = -1 + 34 - 30$

$z = 3$

15.  $y = 1$

16.  $3x + 4(x-2) = x - 5 + 3(2x-1)$

$3x + 4x - 8 = x - 5 + 6x - 3$

$7x - 8 = 7x - 8$

$7x - 7x = 8 - 8$

$0 = 0$

An identity, and the solution is all real numbers.

17.  No solution

18.  $2\{2 - x - (2x - 5)\} = 11 - x$

$2\{2 - x - 2x + 5\} = 11 - x$

$2\{7 - 3x\} = 11 - x$

$14 - 6x = 11 - x$

$14 - 11 = 6x - x$

$3 = 5x$

$\dfrac{3}{5} = \dfrac{\cancel{5}x}{\cancel{5}}$

$\dfrac{3}{5} = x$

# UNIT 4

1.  $x = 5$

2.  $$\frac{x-3}{2} = \frac{2x+4}{5}$$

    $5(x - 3) = 2(2x + 4)$   Cross-multiply.

    $5x - 15 = 4x + 8$

    $5x - 4x = 8 + 15$

    $\quad\quad x = 23$

    It is not necessary to check this, as we did not have a fractional equation—the variable was not in any denominator.

3.  $x = 0$

4.  $$\frac{3x-3}{x-1} = 2$$

    $3x - 3 = 2(x - 1)$   Cross-multiply.

    $3x - 3 = 2x - 2$

    $3x - 2x = -2 + 3$

    $\quad\quad x = 1$

    Check: $\dfrac{3-3}{1-1} \overset{?}{=} 2$

    But we cannot have 0 in the denominator; therefore there is no solution to this equation.

5.  $x = 12$

6.  $$\frac{3}{x} = \frac{4}{x-2}$$

    $3(x - 2) = 4x$

    $\quad 3x - 6 = 4x$

    $3x - 4x = 6$

    $\quad\quad -x = 6$

    $\quad\quad\; x = -6$

    Check:   $\dfrac{3}{x} = \dfrac{4}{x-2}$

    $\qquad\quad \dfrac{3}{(-6)} \overset{?}{=} \dfrac{4}{(-6)-2}$

    $\qquad\quad -\dfrac{1}{2} = -\dfrac{1}{2}$

    True; therefore the solution is $x = -6$.

7. $x = 5$

8. $\dfrac{5 - 2x}{x - 1} = -2$

$5 - 2x = -2(x - 1)$

$5 - 2x = -2x + 2$

$-2x + 2x = 2 - 5$

$0 = -3$

No solution, since 0 does not equal −3.

9. $x = 7$

10. $\dfrac{4}{5}x - \dfrac{1}{4} = -\dfrac{3}{2}x$

$20\left(\dfrac{4}{5}x - \dfrac{1}{4} = -\dfrac{3}{2}x\right)$

$$\dfrac{\overset{4}{\cancel{20}} \cdot 4 \cdot x}{\cancel{5}} - \dfrac{\overset{5}{\cancel{20}} \cdot 1}{\cancel{4}} = -\dfrac{\overset{10}{\cancel{20}} \cdot 3 \cdot x}{\cancel{2}}$$

$16x - 5 = -30x$

$16x + 30x = 5$

$46x = 5$

$\dfrac{\cancel{46}x}{\cancel{46}} = \dfrac{5}{46}$

$x = \dfrac{5}{46}$

The variable was not in any denominator; therefore it is not necessary to check. The solution is $x = \dfrac{5}{46}$.

11. $x = \dfrac{1}{3}$

12.
$$\frac{4}{x-2} - \frac{1}{x} = \frac{5}{x-2}$$

$$\left(\frac{4}{x-2} - \frac{1}{x} = \frac{5}{x-2}\right)x(x-2)$$

$$\left(\frac{4}{x-2} - \frac{1}{x} = \frac{5}{x-2}\right)x(x-2)$$

$$\frac{4 \cdot x(x-2)}{x-2} - \frac{1 \cdot x(x-2)}{x} = \frac{5 \cdot x(x-2)}{x-2}$$

$$4x - (x-2) = 5x$$

$$4x - x + 2 = 5x$$

$$3x + 2 = 5x$$

$$2 = 5x - 3x$$

$$2 = 2x$$

$$\frac{2}{2} = \frac{2x}{2}$$

$$x = 1$$

Check: $\dfrac{4}{x-2} - \dfrac{1}{x} = \dfrac{5}{x-2}$

$$\frac{4}{1-2} - \frac{1}{1} \overset{?}{=} \frac{5}{1-2}$$

$$\frac{4}{-1} - 1 \overset{?}{=} \frac{5}{-1}$$

$$-4 - 1 \overset{?}{=} -5$$

$$-5 = -5$$

True; therefore the solution is $x = 1$.

---

# UNIT 5

1. $a = \dfrac{3cy}{2mx}$

2.
$$2cy + 4d = 3ax - 4b$$
$$2cy + 4d + 4b = 3ax$$
$$\frac{2cy + 4d + 4b}{3x} = \frac{3ax}{3x}$$
$$\frac{2cy + 4d + 4b}{3x} = a$$

3. $a = \dfrac{bx + 7c}{x + 3}$

4.
$$4x + 5c - 2a = 0$$
$$4x + 5c = 2a$$
$$\frac{4x + 5c}{2} = \frac{2a}{2}$$
$$\frac{4x + 5c}{2} = a$$

5. $a = \dfrac{\pi - cy}{x + 2}$

6.　$\dfrac{2ax}{3c} = \dfrac{y}{m}$

　　$2amx = 3cy$

　　$\dfrac{\cancel{2amx}}{\cancel{2am}} = \dfrac{3cy}{2am}$

　　　$x = \dfrac{3cy}{2am}$

7.　$\dfrac{2cy + 4d + 4b}{3a} = x$

8.　$ax + 3a = bx + 7c$

　　$ax - bx = 7c - 3a$

　　$x(a - b) = 7c - 3a$

　　$\dfrac{x(\cancel{a-b})}{\cancel{a-b}} = \dfrac{7c - 3a}{a - b}$

　　　$x = \dfrac{7c - 3a}{a - b}$

9.　$x = \dfrac{2a - 5c}{4}$

10.　$a(x + 2) = \pi - cy$

　　$ax + 2a = \pi - cy$

　　　$ax = \pi - cy - 2a$

　　$\dfrac{\cancel{a}x}{\cancel{a}} = \dfrac{\pi - cy - 2a}{a}$

　　　$x = \dfrac{\pi - cy - 2a}{a}$

11.　$c = \dfrac{2amx}{3y}$

12.　$2cy + 4d = 3ax - 4b$

　　$2cy = 3ax - 4b - 4d$

　　$\dfrac{\cancel{2cy}}{\cancel{2y}} = \dfrac{3ax - 4b - 4d}{2y}$

　　　$c = \dfrac{3ax - 4b - 4d}{2y}$

13.　$c = \dfrac{ax + 3a - bx}{7}$

14.　$4x + 5c - 2a = 0$

　　　$5c = 2a - 4x$

　　　$\dfrac{\cancel{5}c}{\cancel{5}} = \dfrac{2a - 4x}{5}$

　　　　$c = \dfrac{2a - 4x}{5}$

15.　$c = \dfrac{\pi - ax - 2a}{y}$

16.　$C = 2\pi r$

　　$\dfrac{C}{2\pi} = \dfrac{\cancel{2\pi}r}{\cancel{2\pi}}$

　　$\dfrac{C}{2\pi} = r$

17.　$I = \dfrac{V}{R}$

18.　　$A = \dfrac{1}{2}(a + b)h$

　　$2A = 2 \cdot \dfrac{1}{2}(a + b)h$

　　$2A = (a + b)h$

　　$2A = ah + bh$

　$2A - bh = ah$

　$\dfrac{2A - bh}{h} = \dfrac{a\cancel{h}}{\cancel{h}}$

　$\dfrac{2A - bh}{h} = a$

---

# UNIT 6

1.　$x - 5$

2.　$3x + 8$

3.　$8x - 10$

4.　$\dfrac{x}{3}$

5. $2x - 5 = 11$

6. $7x = 35$

7. $x + 20 = 32$

8. $x + 12 = 20$

9. $15 + 2x = 47$

10. $4 + 3x = 17$

11. Jack is 17 and George is 25.

12.         Let $x$ = Length of shorter piece of rope.
Then $x + 10$ = length of longer piece.
The length of the rope is 36 feet.
$$x + (x + 10) = 36$$
$$2x = 26$$
$$x = 13$$
$$x + 10 = 23$$
The shorter piece of rope is 13 feet long, and the other is 23 feet long.

13. The numbers on the raffle tickets are 337, 338, 339, and 340.

14. Let $x$ represent the first even integer.
Then $x + 2$ represents the second consecutive even integer.
And $x + 4$ represents the third consecutive even integer.
The sum of three consecutive even integers is 6480.
$$x + (x + 2) + (x + 4) = 6480$$
$$3x + 6 = 6480$$
$$3x = 6474$$
$$x = 2158$$
$$x + 2 = 2160$$
$$x + 4 = 2162$$
The three consecutive even integers are 2158, 2160, and 2162.

15. The dimensions of the basketball court are 50 ft by 84 ft.

16. Let $x$ represent the width of the parking lot.
Then $3x$ represents the length since it is 3 times the width.
The perimeter is given at 4100 ft.
$$P = 2l + 2w$$
$$4100 = 2(3x) + 2(x)$$
$$4100 = 6x + 2x$$
$$4100 = 8x$$
$$512.5 = x$$
$$3x = 1537.5$$
The width of the parking lot is 512.5 ft, and the length is 1537.5 ft.

17. Aurora and Chris will meet in 0.77 hour or in approximately 46 minutes.

18. Let $x$ represent Rosalie's running time.

Then Sally's running time will be $x - 0.5$ because she started running $\frac{1}{2}$ hour later. The completed chart is:

|  | rate | × | time | = | distance |
|---|---|---|---|---|---|
| Rosalie | 5 mph |  | $x$ |  | $5x$ |
| Sally | 8 mph |  | $x - 0.5$ |  | $8(x - 0.5)$ |

In this problem the distances run by each person are equal.

$$6x = 8(x - 0.5)$$
$$6x = 8x - 4$$
$$4 = 8x - 6x$$
$$4 = 2x$$
$$x = 2$$
$$x - 0.5 = 1.5$$

Sally will reach Rosalie in $1\frac{1}{2}$ hours or 90 minutes.

---

## UNIT 7

1. $(3y)^2 = 3^2 y^2 = 9y^2$

2. $3x^0 = 3 \cdot 1 = 3$

3. $x^2(x^3)^4 = x^2 \cdot x^{12} = x^{2+12} = x^{14}$

4. $(x^2 y^3 z)^4 = x^8 y^{12} z^4$

5. $\left( \dfrac{x^2}{wz} \right)^3 = \dfrac{x^6}{w^3 z^3}$

6. $(2ab)b^2 = 2ab \cdot b^2 = 2 \cdot a \cdot b^{1+2} = 2ab^3$

7. $\dfrac{(x^2 y^3)^2}{5} = \dfrac{x^4 y^6}{5}$

8. $5(x^2 z)^2 = 5 \cdot x^4 z^2 = 5x^4 z^2$

9. $(5x^2 z)^2 = 5^2 x^4 z^2 = 25x^4 z^2$

10. $\left( \dfrac{a^2 b^3 cd^5}{3x^2 w^0} \right)^7 = \dfrac{a^{14} b^{21} c^7 d^{35}}{3^7 x^{14}}$

11. $\dfrac{(2ab)^2}{(3x^3)^2} = \dfrac{2^2 a^2 b^2}{3^2 x^6} = \dfrac{4a^2 b^2}{9x^6}$

12. $(3x^5)^2(2x^3)^3 = 3^2x^{10} \cdot 2^3x^9 = 9x^{10} \cdot 8x^9 = 72x^{19}$

13. $(x^2y)(xy^2) = x^{2+1} \cdot y^{1+2} = x^3y^3$

14. $2(3ab^2)^2 = 2 \cdot 3^2a^2b^4 = 18a^2b^4$

15. $(-4c)^2 = (-4)^2c^2 = 16c^2$

16. $\left(\dfrac{xyz^2}{5a}\right)^3 = \dfrac{x^3y^3z^6}{5^3a^3} = \dfrac{x^3y^3z^6}{125a^3}$

17. $(-2abc)(bcd)(3abc^2) = -2 \cdot 3a^{1+1} \cdot b^{1+1+1} \cdot c^{1+1+2} \cdot d = -6a^2b^3c^4d$

18. $(-2x^2yz)(-5xz)^2(xyz^2)^3 = 2x^2yz \cdot (-5)^2x^2z^2 \cdot x^3y^3z^6$
$$= 2 \cdot 25x^{2+2+3} \cdot y^{1+3} \cdot z^{1+2+6}$$
$$= 50x^7y^4z^9$$

# UNIT 8

1. $\dfrac{a^{-3}}{a^2} = \dfrac{1}{a^3a^2} = \dfrac{1}{a^5}$

2. $\dfrac{a^{-2}x^3}{y^{-1}} = \dfrac{x^3y}{a^2}$ ✓

3. $(x^2y)^{-2} = x^{-4}y^{-2} = \dfrac{1}{x^4y^2}$ ✓

4. $x^5 \cdot x^0 \cdot z^{-7} = \dfrac{x^5 \cdot 1}{z^7} = \dfrac{x^5}{z^7}$ ✗

5. $\dfrac{x^{-2}}{y^{-3}} = \dfrac{y^3}{x^2}$ ✗

6. $-2x^6y^0 = -2x^6 \cdot 1 = -2x^6$ ✗

7. $(3x^{-6}y^5)^{-2} = 3^{-2}x^{12}y^{-10} = \dfrac{x^{12}}{3^2y^{10}} = \dfrac{x^{12}}{9y^{10}}$ ✗

8. $\dfrac{(ab^2)^{-3}}{(x^2y^{-3})^4} = \dfrac{a^{-3}b^{-6}}{x^8y^{-12}} = \dfrac{y^{12}}{a^3b^6x^8}$ ✗

9. $\dfrac{(3ab^5)^{-3}}{2x^{-5}} = \dfrac{3^{-3}a^{-3}b^{-15}}{2x^{-5}} = \dfrac{x^5}{2 \cdot 3^3 a^3b^{15}} = \dfrac{x^5}{54a^3b^{15}}$ ✗

10. $\dfrac{7x^{-1}}{y^2} = \dfrac{7}{xy^2}$ ✗

11. $(5w^{-2})^2(2w^{-2}) = 5^2w^{-4} \cdot 2w^{-2} = \dfrac{5^2 \cdot 2}{w^4 \cdot w^2} = \dfrac{50}{w^6}$

12. $\dfrac{x^{-2}y^{-3}}{(c)^{-2}} = \dfrac{x^{-2}y^{-3}}{c^{-2}} = \dfrac{c^2}{x^2y^3}$

13. $\dfrac{(5a^2b^3)^2}{(-2x)^{-3}} = \dfrac{5^2a^4b^6}{(-2)^{-3}x^{-3}} = (-2)^3 5^2 a^4 b^6 x^3 = -200a^4b^6x^3$

14. $\dfrac{16w^{-1}y^2z^{-3}}{2x} = \dfrac{16y^2}{2wxz^3} = \dfrac{8y^2}{wxz^3}$

15. $\left[\dfrac{b^2}{(a^2b)^{-2}}\right]^{-1} = \left[\dfrac{b^2}{a^{-4}b^{-2}}\right]^{-1} = \dfrac{b^{-2}}{a^4b^2} = \dfrac{1}{a^4b^2 \cdot b^2} = \dfrac{1}{a^4b^4}$

---

# UNIT 9

1. $\dfrac{5a^7b^2}{ab^{10}} = \dfrac{5a^6}{b^8}$

2. $w^5 \cdot w^0 \cdot w^{-7} = \dfrac{w^5 \cdot 1}{w^7} = \dfrac{w^5}{w^7} = \dfrac{1}{w^{7-5}} = \dfrac{1}{w^2}$

3. $(3a^4b^{-2})(a^5b^{-3}) = \dfrac{3a^9}{b^5}$

4. $(4x^{-3}y^7)(-2x^2y^2) = \dfrac{4y^7 \cdot (-2)x^2y^2}{x^3} = \dfrac{-8y^7y^2}{x^{3-2}} = \dfrac{-8y^{7+2}}{x} = \dfrac{-8y^9}{x}$

5. $\dfrac{x^{-4}}{x^4} = \dfrac{1}{x^8}$ ✓

6. $\dfrac{15x^5y^3}{3x^2y^7} = \dfrac{15\cancel{x^5}\cancel{y^3}}{3\cancel{x^2}\cancel{y^7}} = \dfrac{5x^{5-2}}{y^{7-3}} = \dfrac{5x^3}{y^4}$

7. $\dfrac{x^5 \cdot x^{-4}}{x^{-3}} = x^4$ ✓

8. $x(3x^2y^{-3})^2 = x \cdot 3^2 x^4 y^{-6} = \dfrac{9x \cdot x^4}{y^6} = \dfrac{9x^5}{y^6}$

9. $(2w^{-2})^2(5w^{-2}) = \dfrac{20}{w^6}$

10. $x(5xy^{-2})^{-2} = x \cdot 5^{-2}x^{-2}y^{+4} = \dfrac{\cancel{x}y^4}{5^2\cancel{x^2}} = \dfrac{y^4}{25x^{2-1}} = \dfrac{y^4}{25x}$

11. $\dfrac{m^{-9}s^{-8}}{m^{-4}s^3} = \dfrac{1}{m^5s^{11}}$

12. $\dfrac{(3x^2y)^{-1}}{2xy^{-5}} = \dfrac{3^{-1}x^{-2}y^{-1}}{2xy^{-5}} = \dfrac{y^{-5}}{2x \cdot 3x^2y} = \dfrac{y^4}{6x^3}$

13. $\dfrac{(3xy^{-2})^{-3}}{x} = \dfrac{y^6}{27x^4}$

14. $\left[\dfrac{(x^2y)^3}{3x^7y^9}\right]^2 = \left[\dfrac{x^6y^3}{3x^7y^9}\right]^2 = \dfrac{x^{12}y^6}{3^2x^{14}y^{18}} = \dfrac{1}{9x^{14-12}y^{18-6}} = \dfrac{1}{9x^2y^{12}}$

15. $\left[\dfrac{(xy)^2}{x^{-1}}\right]^3 = x^9y^6$

16. $\left[\dfrac{(ab)^{-1}}{(a^{-2}b^3)^3}\right]^{-1} = \left[\dfrac{a^{-1}b^{-1}}{a^{-6}b^9}\right]^{-1} = \dfrac{ab}{a^6b^{-9}} = \dfrac{abb^9}{a^6} = \dfrac{b^{1+9}}{a^{6-1}} = \dfrac{b^{10}}{a^5}$

## UNIT 10

1. $\dfrac{3}{11}$

2. $\dfrac{7}{10} - \dfrac{9}{10} = \dfrac{7-9}{10} = \dfrac{-2}{10} = \dfrac{-1}{5}$

3. $\dfrac{44}{45}$

4. $\dfrac{1}{x} + 5 = \dfrac{1}{x} + \dfrac{5}{1} = \dfrac{1+5x}{x}$

5. $\dfrac{3w-5}{w}$

6. $7 + \dfrac{1}{x} = \dfrac{7}{1} + \dfrac{1}{x} = \dfrac{7x+1}{x}$

7. $\dfrac{29}{90}$

8. $\dfrac{11}{t} + \dfrac{7}{r} = \dfrac{11r+7t}{rt}$

9. $\dfrac{2}{x}$

10. $\dfrac{10}{x+1} + \dfrac{3}{x+1} = \dfrac{10+3}{x+1} = \dfrac{13}{x+1}$

11. $\dfrac{1}{a}$

12. $\dfrac{-s}{9} + \dfrac{k}{10} = \dfrac{-10s+9k}{90}$

13. $\dfrac{3x}{10}$

14. $\dfrac{x+1}{2} - \dfrac{3}{5} = \dfrac{5(x+1)-6}{10}$

$= \dfrac{5x+5-6}{10}$

$= \dfrac{5x-1}{10}$

15. $\dfrac{5x+1}{6}$

16. $\dfrac{x+2}{2} - \dfrac{x+3}{3} = \dfrac{3(x+2)-2(x+3)}{6}$

$= \dfrac{3x+6-2x-6}{6}$

$= \dfrac{x}{6}$

# UNIT 11

1.  $\dfrac{5}{3}$

2.  $\dfrac{5}{18} \div \dfrac{3}{14} = \dfrac{5}{\overset{}{\underset{9}{\cancel{18}}}} \cdot \dfrac{\overset{7}{\cancel{14}}}{3} = \dfrac{35}{27}$

3.  $\dfrac{-2}{3}$

4.  $\dfrac{9}{14} \div \dfrac{5}{21} = \dfrac{9}{\overset{}{\underset{2}{\cancel{14}}}} \cdot \dfrac{\overset{3}{\cancel{21}}}{5} = \dfrac{27}{10}$

5.  $\dfrac{14}{27}$

6.  $\left(\dfrac{2}{x} \cdot \dfrac{x}{5}\right) \div w = \left(\dfrac{2}{\cancel{x}} \cdot \dfrac{\cancel{x}}{5}\right) \div w = \dfrac{2}{5} \cdot \dfrac{1}{w} = \dfrac{2 \cdot 1}{5 \cdot w} = \dfrac{2}{5w}$

7.  3

8.  $\dfrac{3 - \dfrac{2}{5}}{3 + \dfrac{2}{5}} = \left(\dfrac{3}{1} - \dfrac{2}{5}\right) \div \left(\dfrac{3}{1} + \dfrac{2}{5}\right)$

    $= \dfrac{15 - 2}{5} \div \dfrac{15 + 2}{5}$

    $= \dfrac{13}{5} \div \dfrac{17}{5}$

    $= \dfrac{13}{\cancel{5}} \cdot \dfrac{\cancel{5}}{17}$

    $= \dfrac{13}{17}$

9.  $\dfrac{4}{5}$

10. $\dfrac{\dfrac{a}{2} - \dfrac{3}{5}}{2} = \left(\dfrac{a}{2} - \dfrac{3}{5}\right) \div 2$

    $= \dfrac{5a - 6}{10} \div \dfrac{2}{1}$

    $= \dfrac{5a - 6}{10} \cdot \dfrac{1}{2}$

    $= \dfrac{5a - 6}{20}$

11. $\dfrac{28}{x-1}$

12. $\dfrac{\dfrac{2}{x}-5}{x} = \left(\dfrac{2}{x}-\dfrac{5}{1}\right) \div x$

$$= \dfrac{2-5x}{x}\cdot\dfrac{1}{x}$$

$$= \dfrac{2-5x}{x^2}$$

13. $\dfrac{36-x}{2}$

14. $\dfrac{x+\dfrac{x+2}{2}}{\dfrac{x}{2}} = \left(\dfrac{x}{1}+\dfrac{x+2}{2}\right)\div\dfrac{x}{2}$

$$= \dfrac{2x+x+2}{2}\div\dfrac{x}{2}$$

$$= \dfrac{3x+2}{2}\cdot\dfrac{2}{x}$$

$$= \dfrac{3x+2}{\cancel{2}}\cdot\dfrac{\cancel{2}}{x}$$

$$= \dfrac{3x+2}{x}$$

15. $\dfrac{a+2b}{a+b}$

# UNIT 12

1. $\sqrt{5}\cdot\sqrt{20} = 10$

2. $\sqrt{75} = \sqrt{25\cdot3} = 5\sqrt{3}$

3. $\sqrt{2a}\cdot\sqrt{3b} = \sqrt{6ab}$

4. $\sqrt{0} = 0$, by definition

5. $\sqrt{3}\cdot\sqrt{6} = 3\sqrt{2}$

6. $\sqrt{64t^2} = 8t$

7. $\sqrt{w^4} = w^2$

8. $\sqrt{45} = \sqrt{9\cdot5} = 3\sqrt{5}$

9. $\sqrt{\dfrac{25}{x^2}} = \dfrac{5}{x}$

10. $\dfrac{\sqrt{20}}{\sqrt{5}} = \sqrt{\dfrac{20}{5}} = \sqrt{4} = 2$

11. $11\sqrt{2}+3\sqrt{2} = 14\sqrt{2}$

12. $\sqrt{3x}\cdot\sqrt{3x} = \sqrt{9x^2} = 3x$

13. $7\sqrt{40}-2\sqrt{10} = 12\sqrt{10}$

14. $\sqrt{12y^8} = \sqrt{4y^8\cdot3} = 2y^4\sqrt{3}$

15. $\sqrt{50x^4} = 5x^2\sqrt{2}$

16. $\sqrt{3}\left(\sqrt{2}+1\right) = \sqrt{6}+\sqrt{3}$

17. $\sqrt{5}\left(\sqrt{5}+\sqrt{3}\right) = 5+\sqrt{15}$

18. $\sqrt{2a^2c}\cdot\sqrt{2ac} = \sqrt{4a^3c^2} = \sqrt{4a^2c^2\cdot a}$

$$= 2ac\sqrt{a}$$

# UNIT 13

1.  2

2.  $(-1)^{2/3} = \left(\sqrt[3]{-1}\right)^2 = (-1)^2 = 1$

3.  $\sqrt{-4}$
    No solution; we cannot take the square root of a negative number in the set of reals.

4.  $4^{3/2} = \left(\sqrt{4}\right)^3 = (2^3) = 8$

5.  2

6.  $4^{-1/2} = \left(\sqrt{4}\right)^{-1} = 2^{-1} = \dfrac{1}{2}$

7.  $\dfrac{1}{\sqrt{x}}$

8.  $x^{1/3} = \sqrt[3]{x}$

9.  $\sqrt[5]{a^2}$

10. $4^{-3/2} = \left(\sqrt{4}\right)^{-3} = 2^{-3} = \dfrac{1}{2^3} = \dfrac{1}{8}$

11. $\sqrt{x+1}$

12. $x^{8/3} = \sqrt[3]{x^8} = \sqrt[3]{x^3 \cdot x^3 \cdot x^2} = x \cdot x\sqrt[3]{x^2} = x^2\sqrt[3]{x^2}$

13. $2\sqrt{x}$

14. $x^{11/2} = \sqrt{x^{11}} = \sqrt{x^{10} \cdot x} = x^5\sqrt{x}$

15. $\dfrac{1}{\sqrt{5x}}$

16. $(18x^3)^{1/2} = \sqrt{18x^3} = \sqrt{9 \cdot 2 \cdot x^2 x} = \sqrt{9x^2 \cdot 2x} = 3x\sqrt{2x}$

17. $\sqrt[3]{4x^2}$

18. $(-64)^{2/3} = \left(\sqrt[3]{-64}\right)^2 = (-4)^2 = 16$

19. $(7x)^{1/2}$

20. $\sqrt[3]{2x} = (2x)^{1/3}$

C21. 1.872

C22.   $\sqrt[4]{17} = (17)^{1/4} = 2.031$

Press  $\boxed{1}\,\boxed{7}\,\boxed{\wedge}\,\boxed{(}\,\boxed{(}\,\boxed{1}\,\boxed{\div}\,\boxed{4}\,\boxed{)}$ $\boxed{\text{ENTER}}$.

## UNIT 14

1.  $\dfrac{y^{2/3}}{y^{1/3}} = y^{2/3 - 1/3} = y^{1/3} = \sqrt[3]{y}$

2.  $\left(y^{3/5}\right)^{1/4} = y^{3/5 \cdot 1/4} = y^{3/20} = \sqrt[20]{y^3}$

3.  $x^{1/2} \cdot x^{2/5} = x^{1/2 + 2/5} = x^{(5+4)/10} = x^{9/10} = \sqrt[10]{x^9}$

4.  $\left(\dfrac{a^4}{c^2}\right)^{1/2} = \dfrac{a^2}{c}$

5.  $\left[\left(\sqrt{4}\right)^{-1}\right]^2 = \left[(2)^{-1}\right]^2 = (2)^{-2} = \dfrac{1}{2^2} = \dfrac{1}{4}$

6.  $\left(\sqrt{x}\right)^{1/2} = \left(x^{1/2}\right)^{1/2} = x^{1/2 \cdot 1/2} = x^{1/4} = \sqrt[4]{x}$

7.  $\left(8x^2\right)^{1/3} = 8^{1/3} x^{2/3} = 2\sqrt[3]{x^2}$

8.  $\left(\dfrac{2^{-3} \cdot 2^5}{2^{-2}}\right)^3 = \dfrac{2^{-9} \cdot 2^{15}}{2^{-6}}$

    $= \dfrac{2^{15} \cdot 2^6}{2^9}$

    $= \dfrac{2^{21}}{2^9}$

    $= 2^{21-9}$

    $= 2^{12}$

An alternative, shorter approach would be

$\left(\dfrac{2^{-3} \cdot 2^5}{2^{-2}}\right)^3 = \left(\dfrac{2^2}{2^{-2}}\right)^3$

    $= \left(2^2 \cdot 2^2\right)^3$

    $= \left(2^4\right)^3$

    $= 2^{12}$

9.  $\left(\dfrac{x^{1/3}}{x^{2/3}}\right)^3 = \dfrac{x}{x^2} = \dfrac{1}{x}$

10.  $x^{1/2} \cdot x^{5/2} = x^{1/2 + 5/2} = x^{6/2} = x^3$

11.  $\left(\sqrt[3]{x^2}\right)^{1/2} = \left(x^{2/3}\right)^{1/2} = x^{1/3} = \sqrt[3]{x}$

12.  $\dfrac{x^{-7/2} \cdot x^{3/2}}{\sqrt{x} \cdot x^{-3/2}} = \dfrac{x^{-7/2} \cdot x^{3/2}}{x^{1/2} \cdot x^{-3/2}}$

    $= \dfrac{x^{3/2} \cdot x^{3/2}}{x^{7/2} \cdot x^{1/2}}$

    $= \dfrac{x^{3/2 + 3/2}}{x^{7/2 + 1/2}}$

    $= \dfrac{x^{6/2}}{x^{8/2}}$

    $= \dfrac{x^3}{x^4}$

    $= \dfrac{1}{x}$

13.  $\left(8\sqrt{x}\right)^{-2/3} = \left(8x^{1/2}\right)^{-2/3}$

    $= 8^{-2/3} x^{-1/3}$

    $= \dfrac{1}{8^{2/3} x^{1/3}}$

    $= \dfrac{1}{\left(\sqrt[3]{8}\right)^2 \sqrt[3]{x}}$

    $= \dfrac{1}{4\sqrt[3]{x}}$

14.  $\left(\dfrac{27^{5/3} \cdot 27^{-1/3}}{27^{1/3}}\right)^2 = \dfrac{27^{10/3} \cdot 27^{-2/3}}{27^{2/3}}$

    $= \dfrac{27^{10/3}}{27^{2/3} \cdot 27^{2/3}}$

    $= \dfrac{27^{10/3}}{27^{2/3 + 2/3}}$

    $= \dfrac{27^{10/3}}{27^{4/3}}$

    $= 27^{10/3 - 4/3}$

    $= 27^{6/3}$

    $= 27^2$

15. 

$$\left(\frac{3x^{-1}}{\sqrt{x}}\right)^2 = \left(\frac{3x^{-1}}{x^{1/2}}\right)^2$$

$$= \frac{3^2 x^{-2}}{x}$$

$$= \frac{9}{x \cdot x^2}$$

$$= \frac{9}{x^3}$$

## UNIT 15

1. $\dfrac{1}{\left(x^2+1\right)^2}$

2. $x^{-1}+2^{-2} = \dfrac{1}{x}+\dfrac{1}{2^2}$

$$= \frac{1}{x} + \frac{1}{4}$$

$$= \frac{4+x}{4x}$$

3. $\dfrac{1-x}{x}$

4. $\dfrac{1}{x^{-1}+y^{-1}} = \dfrac{1}{\dfrac{1}{x}+\dfrac{1}{y}}$

$$= 1 \div \left(\frac{1}{x} + \frac{1}{y}\right)$$

$$= 1 \div \left(\frac{y+x}{xy}\right)$$

$$= 1 \cdot \frac{xy}{y+x}$$

$$= \frac{xy}{y+x}$$

5. $\dfrac{4}{9}$

6. $x^{-1}+2y^{-1} = \dfrac{1}{x}+\dfrac{2}{y}$  ◄── Be careful that the 2 is in the numerator.

$$= \left(\frac{1}{x} + \frac{2}{y}\right)$$

$$= \frac{y+2x}{xy}$$

7. $\dfrac{ab+1}{b}$

8. $5(x+y)^{-1} = \dfrac{5}{(x+y)}$

9. $\dfrac{(2x-3)^2}{x}$

10. $3x^{-2}+y = \dfrac{3}{x^2}+y$

$$= \frac{3}{x^2} + \frac{y}{1}$$

$$= \frac{3+x^2 y}{x^2}$$

11. $\dfrac{a^2-1}{a^2+1}$

12. $\dfrac{3^{-1}+2^{-1}}{3^{-1}-2^{-1}} = \dfrac{\dfrac{1}{3}+\dfrac{1}{2}}{\dfrac{1}{3}-\dfrac{1}{2}}$

$$= \left(\frac{1}{3}+\frac{1}{2}\right) \div \left(\frac{1}{3}-\frac{1}{2}\right)$$

$$= \frac{2+3}{6} \div \frac{2-3}{6}$$

$$= \frac{5}{6} \div \frac{-1}{6}$$

$$= \frac{5}{6} \cdot \frac{6}{-1}$$

$$= -5$$

13. $\dfrac{ab+1}{ab^2}$

14. $\dfrac{3ab}{a^{-1}+b} = \dfrac{3ab}{\dfrac{1}{a}+b}$

$= 3ab \div \left(\dfrac{1}{a}+b\right)$

$= 3ab \div \left(\dfrac{1}{a}+\dfrac{b}{1}\right)$

$= 3ab \div \left(\dfrac{1}{a}\searrow\dfrac{b}{1}\right)$

$= 3ab \div \left(\dfrac{1+ab}{a}\right)$

$= 3ab \cdot \left(\dfrac{a}{1+ab}\right)$

$= \dfrac{3ab}{1} \cdot \dfrac{a}{1+ab}$

$= \dfrac{3a^2b}{1+ab}$

# UNIT 16

1. $10c^3x^2 - 2c^2x^2 - 6cx^3$

2. $(x+4)(x+5) = x^2 + 9x + 20$

3. $x^2 - 9x + 14$

4. $(x-1)(x-5) = x^2 - 6x + 5$

5. $x^2 - x - 6$

6. $(a+5)^2 = a^2 + 2(5a) + 25$
   $\phantom{(a+5)^2} = a^2 + 10a + 25$

7. $x^2 - 4$

8. $(x-1)^2 = x^2 - 2x + 1$

9. $x^2 - x - 12$

10. $(2x+1)(x+1) = 2x(x+1) + (x+1)$
    $\phantom{(2x+1)(x+1)} = 2x^2 + 2x + x + 1$
    $\phantom{(2x+1)(x+1)} = 2x^2 + 3x + 1$

11. $2x^2 + 3x - 20$

12. $(3x-2)(x+7) = 3x(x+7) - 2(x+7)$
    $\phantom{(3x-2)(x+7)} = 3x^2 + 21x - 2x - 14$
    $\phantom{(3x-2)(x+7)} = 3x^2 + 19x - 14$

13. $5x^2 + 11x + 2$

14. $(3x+1)(3x-1) = 3x(3x-1) + (3x-1)$
    $\phantom{(3x+1)(3x-1)} = 9x^2 - 3x + 3x - 1$
    $\phantom{(3x+1)(3x-1)} = 9x^2 - 1$

15. $8x^2 + 14x + 3$

16. $(5x - 1)(x + 2) = 5x(x + 2) - (x + 2)$
$$= 5x^2 + 10x - x - 2$$
$$= 5x^2 + 9x - 2$$

17. $4x^2 + 12x + 9$

18. $(4x - 3)^2 = 16x^2 + 2(4x)(-3) + 9$
$$= 16x^2 - 24x + 9$$

19. $6a^2 - ab - b^2$

20. $(x + y)(x - y) = x(x - y) + y(x - y)$
$$= x^2 - xy + xy - y^2$$
$$= x^2 - y^2$$

21. $6x^2 - 13x - 5$

22. $x(x - 4)^2 = x(x^2 - 8x + 16)$
$$= x^3 - 8x^2 + 16x$$

23. $x^3 - 10x^2 + 25x$

24. $(x - 2)(x^3 - 4x^2 + 7x - 1) = x(x^3 - 4x^2 + 7x - 1) - 2(x^3 - 4x^2 + 7x - 1)$
$$= x^4 - 4x^3 + 7x^2 - x - 2x^3 + 8x^2 - 14x + 2$$
$$= x^4 - 6x^3 + 15x^2 - 15x + 2$$

25. $x^4 - 2x^2 - 3$

26. $(x + 2y)(x - 3y) = x(x - 3y) + 2y(x - 3y)$
$$= x^2 - 3xy + 2xy - 6y^2$$
$$= x^2 - xy - 6y^2$$

27. $-2a^2 + 7a - 3$

28. $(x^2 - 3x + 1)(x^3 - 2x) = x^2(x^3 - 2x) - 3x(x^3 - 2x) + 1(x^3 - 2x)$
$$= x^5 - 2x^3 - 3x^4 + 6x^2 + x^3 - 2x$$
$$= x^5 - 3x^4 - x^3 + 6x^2 - 2x$$

29. $x^3 - 3x^2 + 5x - 15$

30. $(5a - 3b)(-2a + 6b) = 5a(-2a + 6b) - 3b(-2a + 6b)$
$$= -10a^2 + 30ab + 6ab - 18b^2$$
$$= -10a^2 + 36ab - 18b^2$$

## UNIT 17

1. $x + 3$

2. $$12 + \frac{49}{x - 4}$$
$$x - 4 \overline{)12x + \phantom{0}1}$$
$$\underline{12x - 48}$$
$$49$$

3. $x + 3 + \dfrac{1}{x + 5}$

4. $$2x - 3 + \frac{4}{5x + 1}$$
$$5x + 1 \overline{)10x^2 - 13x + 1}$$
$$\underline{10x^2 + 2x}$$
$$-15x + 1$$
$$\underline{-15x - 3}$$
$$4$$

5. $x - 4$

6.
$$3x - 8 + \frac{2}{2x-1}$$
$$2x-1\overline{)6x^2 - 19x + 10}$$
$$\underline{6x^2 - 3x}$$
$$-16x + 10$$
$$\underline{-16x + 8}$$
$$2$$

9.  $x^2 - x + 1$

10.
$$x^2 - 2x + 4 + \frac{-6}{x+2}$$
$$x+2\overline{)x^3 + 0 \quad + 0 \quad + 2}$$
$$\underline{x^3 + 2x^2}$$
$$-2x^2 + 0$$
$$\underline{-2x^2 - 4x}$$
$$4x + 2$$
$$\underline{4x + 8}$$
$$-6$$

7.  $2x + 7$

8.
$$5x + 5 + \frac{4}{x-1}$$
$$x-1\overline{)5x^2 + 0 - 1}$$
$$\underline{5x^2 - 5x}$$
$$+5x - 1$$
$$\underline{5x - 5}$$
$$4$$

# UNIT 18

1.  $(x + 3)(x - 1)$
2.  $(x - 8)(x - 7)$
3.  $x(x + 1)$
4.  $3xy(x - 4y)$
5.  $3b(x^2 + 9b)$
6.  $(x - 3)(x + 2)$
7.  $(x + 3)(x + 2)$
8.  $(x - 3)(x - 4)$
9.  Prime
10.  $(x + 2)(x - 4)$
11.  Prime
12.  $2x(x^2 + x + 11)$
13.  $5(x^2 - x - 1)$
14.  $(x + 5)(x - 2)$
15.  $(x + 10)(x - 3)$
16.  $(x + 6)(x + 1)$
17.  $(x + 1)(x + 1)$
18.  $(x - 3)(x - 3)$
19.  $(x - 8)(x + 7)$
20.  $(x + 9)(x - 5)$
21.  $(x + 8)(x + 8)$
22.  $(x - 5)(x - 8)$
23.  $(x + 9)(x - 2)$
24.  $x(x^2 + x + 5)$
25.  $(x - 7)(x - 3)$
26.  $(x - 9)(x + 2)$

# UNIT 19

1.  $(x + 6)(x - 6)$
2.  $w^2 - 81 = (w + 9)(w - 9)$
3.  $(x + y)(x - y)$
4.  $16x^2 - 9 = (4x + 3)(4x - 3)$

5. $3(b + 5)(b - 5)$

6. $9 - 6x + x^2 = x^2 - 6x + 9$
$= (x - 3)(x - 3)$

7. $2(x + 1)(x - 1)$

8. $3abc^2 - 3abd^2 = 3ab(c^2 - d^2)$
$= 3ab(c + d)(c - d)$

9. $(x + 4)(x - 2)$

10. $x^2 - x + 7$    prime

11. $x(x + 6)(x - 6)$

12. Prime, not the difference of two squares

13. $(x + 10)(x + 3)$

14. $3r^3 - 6r^2 - 45r = 3r(r^2 - 2r - 15)$
$= 3r(r - 5)(r + 3)$

15. $(x + 7)(x - 2)$

16. $2a^2b^2c^2 - 4ab^2c^2 + 2b^2c^2 = 2b^2c^2(a^2 - 2a + 1)$
$= 2b^2c^2(a - 1)(a - 1)$

17. $5y(x^2 - 3x - 2)$

18. $5x^4 + 10x^3 - 15x^2 = 5x^2(x^2 + 2x - 3)$
$= 5x^2(x + 3)(x - 1)$

19. $3(x - 2)(x + 2)$

20. $a^2b^2 - a^2c^2 = a^2(b^2 - c^2)$
$= a^2(b + c)(b - c)$

21. $2x(y - 2)(y - 25)$

22. $10ab^2 - 140ab + 330a = 10a(b^2 - 14b + 33)$
$= 10a(b - 11)(b - 3)$

23. $w^2x^2(y + 9)(y - 2)$

24. $2ax^2 - 2ax - 40a = 2a(x^2 - x - 20)$
$= 2a(x - 5)(x + 4)$

25. $(2a + 3b)(2a - 3b)$

26. $2x^2 - 10x - 12 = 2(x^2 - 5x - 6)$
$$= 2(x - 6)(x + 1)$$

27. $4rs^2(r - 3)(r - 9)$

28. $2y^2z + 38yz + 96z = 2z(y^2 + 19y + 48)$
$$= 2z(y + 3)(y + 16)$$

29. $3a^2b(b^2 + 1)(b + 1)(b - 1)$

30. $a^2x^4 - 81a^2 = a^2(x^4 - 81)$
$$= a^2(x^2 + 9)(x^2 - 9)$$
$$= a^2(x^2 + 9)(x + 3)(x - 3)$$

# UNIT 20

1. $(7x + 3)(x + 1)$

2. $2y^2 + 5y - 3 = 2y^2 + 6y - y - 3$
$$= 2y(y + 3) - (y + 3)$$
$$= (y + 3)(2y - 1)$$

product = 6
sum = 5
2 and 3

3. $(3x + 4)(2x + 1)$

4. $3x^3 - 5x^2 - 9x + 15 = x^2(3x - 5) - 3(3x - 5)$
$$= (3x - 5)(x^2 - 3)$$

5. $4(x - 3)(x - 4)$

6. $6x^2 + 13x + 6 = 6x^2 + 9x + 4x + 6$
$$= 3x(2x + 3) + 2(2x + 3)$$
$$= (2x + 3)(3x + 2)$$

product = 36
sum = 13
4 and 9

7. $2(x + 4)(x - 3)$

8. $4x^3 - 10x^2 - 6x + 9 = 2x^2(2x - 5) - 3(2x - 3)$

The technique has shown that the expression cannot be factored. The expression is prime.

9. $(5x + 1)(x - 1)$

10. $4x^2 + 8x + 4 = 4(x^2 + 2x + 1)$
$$= 4(x + 1)(x + 1)$$

11. $(7x - 1)(x + 2)$

12. $2x^2 - 7x + 6 = 2x^2 - 4x - 3x + 6$
$$= 2x(x - 2) - 3(x - 2)$$
$$= (x - 2)(2x - 3)$$

product = 12
sum    = 7
3 and 4

13. $(2y - 7)(y - 5)$

14. $7x^2 + 32x - 15 = 7x^2 + 35x - 3x - 15$
$$= 7x(x + 5) - 3(x + 5)$$
$$= (x + 5)(7x - 3)$$

product   = 105
difference = 32
3 and 35

15. $3z(3x + 1)(3x - 1)$

16. $6z^2 + 2z - 4 = 2(3z^2 + z - 2)$
$$= 2[3z^2 + 3z - 2z - 2]$$
$$= 2[3z(z + 1) - 2(z + 1)]$$
$$= 2[(z + 1)(3z - 2)]$$
$$= 2(z + 1)(3z - 2)$$

product   = 6
difference = 1
2 and 3

17. $z(x - 8)(x - 8)$

18. $6xw^2 + 16wx - 6x = 2x(3w^2 + 8w - 3)$
$$= 2x(3w^2 + 9w - w - 3)$$
$$= 2x[3w(w + 3) - (w + 3)]$$
$$= 2x[(w + 3)(3w - 1)]$$
$$= 2x(w + 3)(3w - 1)$$

product   = 9
difference = 8
1 and 9

19. $(2y + x)(4 + x)$

20. $2x^2 + 5x - 2$     prime

product   = 4
difference = 5

21. $(4x - 3)(2x + 9)$

22. $xy^3 + 2y^2 - xy - 2 = y^2(xy + 2) - (xy + 2)$
$$= (xy + 2)(y^2 - 1)$$
$$= (xy + 2)(y + 1)(y - 1)$$

23. $(6x - 5)(2x + 1)$

24. $x^4 - y^4 = (x^2 + y^2)(x^2 - y^2)$
$$= (x^2 + y^2)(x + y)(x - y)$$

25. $(1 + a^2)(1 + a)(1 - a)$

---

# UNIT 21

1. $x = -7$  or  $x = 2$

2. $x^2 + 13x + 30 = 0$
$(x + 10)(x + 3) = 0$
$x + 10 = 0$  or  $x + 3 = 0$
$x = -10$  $x = -3$

3. $x^2 - x + 7 = 0$
Prime. There is no real solution.

4. $4x^2 + 8x + 4 = 0$
$4(x^2 + 2x + 1) = 0$
$4(x + 1)(x + 1) = 0$
$x + 1 = 0$
$x = -1$

5. $x = 0$  or  $x = -5$

6. $x^2 + 2x = 8$
$x^2 + 2x - 8 = 0$
$(x + 4)(x - 2) = 0$
$x + 4 = 0$  or  $x - 2 = 0$
$x = -4$  $x = 2$

7. $x = 0$  $x = -2$  $x = -3$

8. $5x^2 - 5x = 0$
$5x(x - 1) = 0$
$5x = 0$  or  $x - 1 = 0$
$x = 0$  $x = 1$

9. $x = \dfrac{1}{2}$  or  $x = 3$

10. $2x^2 + 8x + 6 = 0$
$2(x^2 + 4x + 3) = 0$
$2(x + 1)(x + 3) = 0$
$x + 1 = 0$  or  $x + 3 = 0$
$x = -1$  $x = -3$

11. $z = 3$  or  $z = -7$

12. $10x - 10 = 19x - x^2$
$x^2 - 9x - 10 = 0$
$(x - 10)(x + 1) = 0$
$x - 10 = 0$  or  $x + 1 = 0$
$x = 10$  $x = -1$

13. $x = 0$  or  $x = \dfrac{-2}{3}$

14. $2 - 2x^2 = 0$
$2 = 2x^2$
$1 = x^2$
$\pm 1 = x$
$x = 1$  or  $x = -1$

15. $w = \dfrac{1}{2}$  or  $w = -4$

16. $x^3 + 3x^2 - 10x = 0$
$x(x^2 + 3x - 10) = 0$
$x(x + 5)(x - 2) = 0$
$x = 0$  $x + 5 = 0$  $x - 2 = 0$
$x = 0$  $x = -5$  $x = 2$

17. $x = -1$      $x = 7$      $x = 3$

19. $x = 0$    $x = -8$

18.
$$2x^3 - x^2 + 14x - 7 = 0$$
$$x^2(2x - 1) + 7(2x - 1) = 0$$
$$(2x - 1)(x^2 + 7) = 0$$

$2x - 1 = 0$    $x^2 + 7 = 0$

     $2x = 1$    no solution to this equation

        $x = \dfrac{1}{2}$

20.
$$12x^2 + 5x - 2 = 0$$
$$(4x - 1)(3x + 2) = 0$$

$4x - 1 = 0$    or    $3x + 2 = 0$

    $4x = 1$           $3x = -2$

    $x = \dfrac{1}{4}$         $x = \dfrac{-2}{3}$

---

# UNIT 22

1.   $x = \dfrac{-3 \pm \sqrt{13}}{2}$

2.   $2x^2 - 3x - 2 = 0$

Case III:

| product = 4 |
| difference = 3 |
| 1 and 4 |

$$2x^2 - 3x - 2 = 0$$
$$2x^2 + x - 4x - 2 = 0$$
$$x(2x + 1) - 2(2x + 1) = 0$$
$$(2x + 1)(x - 2) = 0$$

$2x + 1 = 0$     $x - 2 = 0$

   $2x = -1$       $x = 2$

     $x = -\dfrac{1}{2}$

3.   $x = \dfrac{-1 \pm \sqrt{5}}{2}$

4.   $2x^2 - 3x - 1 = 0$

Thus $a = 2$, $b = -3$, $c = -1$.

$$x = \frac{-b \pm \sqrt{b^2 - 4ac}}{2a}$$

$$= \frac{-(-3) \pm \sqrt{9 - 4(-2)}}{4} \qquad \text{since } ac = 2(-1) = -2$$

$$= \frac{3 \pm \sqrt{9 + 8}}{4}$$

$$= \frac{3 \pm \sqrt{17}}{4}$$

5. $x = 4$     or     $x = -5$

6. $5\left(\dfrac{1}{5}x^2 - 5x + 1 = 0\right)$

   $x^2 - 25x + 5 = 0$

   Thus $a = 1$, $b = -25$, $c = 5$.

   $$x = \frac{-(-25) \pm \sqrt{(-25)^2 - 4(5)}}{2}$$

   $$= \frac{25 \pm \sqrt{625 - 20}}{2}$$

   $$= \frac{25 \pm \sqrt{605}}{2}$$

7. $x = 4$     or     $x = -1$

8. $x^2 - 2x + 2 = 0$

   Also, $a = 1$, $b = -2$, $c = 2$.

   $$x = \frac{-(-2) \pm \sqrt{4 - 4(2)}}{2}$$

   $$= \frac{2 \pm \sqrt{-4}}{2}$$

   There is no real solution. We cannot take the square root of a negative number in the reals.

9. $x = 0$     or     $x = \dfrac{1}{2}$

10. $4x^2 - 4x + 1 = 0$

    Also, $a = 4$, $b = -4$, $c = 1$.

    $$x = \frac{-(-4) \pm \sqrt{16 - 4(4)}}{2(4)}$$

    $$= \frac{4 \pm \sqrt{0}}{8}$$

    $$= \frac{1}{2} \qquad \text{since } \sqrt{0} = 0$$

11. There is no real solution.

12. $x^2 - 2x - 10 = 0$

Also, $a = 1$, $b = -2$, $c = -10$.

$$x = \frac{-(-2) \pm \sqrt{4 - 4(-10)}}{2}$$

$$= \frac{2 \pm \sqrt{44}}{2}$$

$$= \frac{2 \pm 2\sqrt{11}}{2} \qquad \text{since } \sqrt{44} = \sqrt{4 \cdot 11} = 2\sqrt{11}$$

$$= \frac{2(1 \pm \sqrt{11})}{2}$$

$$= 1 \pm \sqrt{11}$$

13. $x = \dfrac{-1 \pm \sqrt{37}}{6}$

14. $2x^2 = \sqrt{3}x - 4 = 0$

Also, $a = 2$, $b = \sqrt{3}$, $c = -4$.

$$x = \frac{-\sqrt{3} \pm \sqrt{3 - 4(-8)}}{4} \qquad \text{since } b^2 = \left(\sqrt{3}\right)^2 = \left(3^{1/2}\right)^2 = 3$$

$$= \frac{-\sqrt{3} \pm \sqrt{35}}{4}$$

15. $x = 0 \qquad x = 4 \qquad x = -3$

16. $2x^4 + 2x^3 + 2x^2 = 0$

$2x^2(x^2 + x + 1) = 0$

$2x^2 = 0 \qquad x^2 + x + 1 = 0$

$x^2 = 0 \qquad$ Use the quadratic formula with $a = 1$, $b = 1$, $c = 1$.

$$x = 0 \qquad\qquad x = \frac{-1 \pm \sqrt{1 - 4}}{2}$$

$$= \frac{-1 \pm \sqrt{-3}}{2}$$

There is no real solution to this part since we cannot have a negative number under the square root symbol in the reals.

Therefore the only solution to this equation within the set of real numbers is $x = 0$.

## UNIT 23

1.

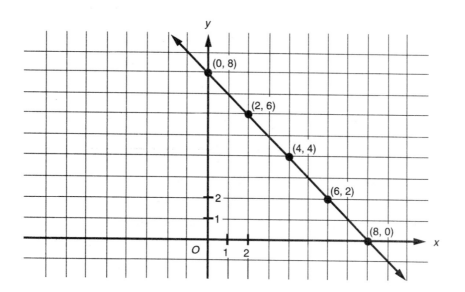

The $x$-intercept is 8.
The $y$-intercept is 8.

2.

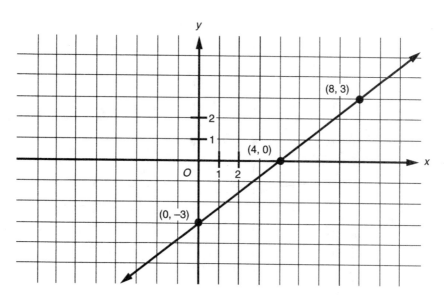

$3x - 4y = 12$

The $x$-intercept is where $y = 0$: $\quad 3x - 4(0) = 12$
$$3x = 12$$
$$x = 4$$

Therefore the $x$-intercept is 4.

The $y$-intercept is where $x = 0$: $\quad 3(0) - 4y = 12$
$$-4y = 12$$
$$y = -3$$

Therefore the $y$-intercept is –3.

3.

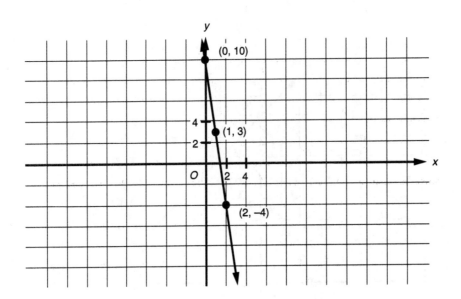

The $x$-intercept is $\dfrac{10}{7}$.

The $y$-intercept is 10.

4.

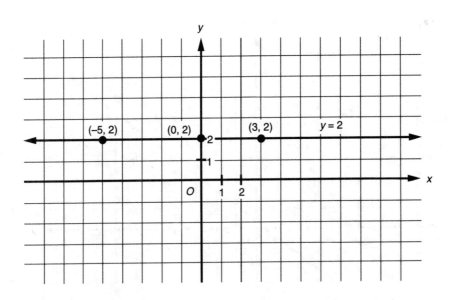

There is no $x$-intercept.

The $y$-intercept is 2.

5.

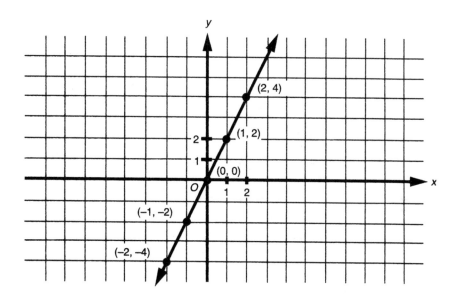

The $x$-intercept is 0.
The $y$-intercept is 0.

6.

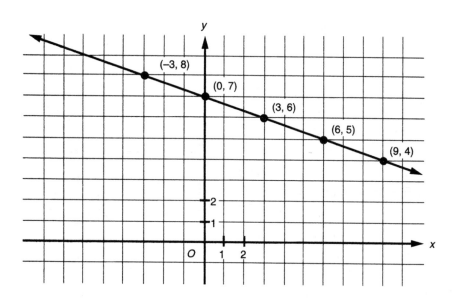

$x + 3y = 21$

$x + 3(0) = 21$

$\quad x = 21;\quad$ the $x$-intercept is 21.

$0 + 3y = 21$

$\quad y = 7;\quad$ the $y$-intercept is 7.

7.

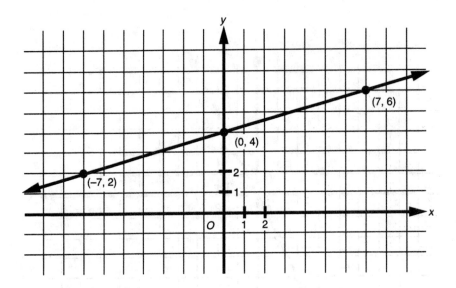

The *x*-intercept is −14.
The *y*-intercept is 4.

8.

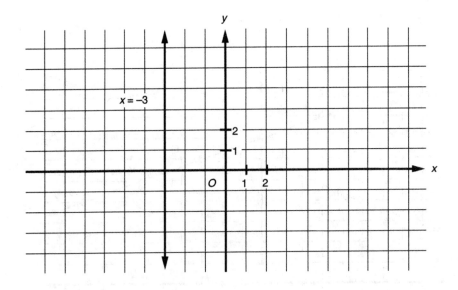

The *x*-intercept is −3.
There is no *y*-intercept.

9.

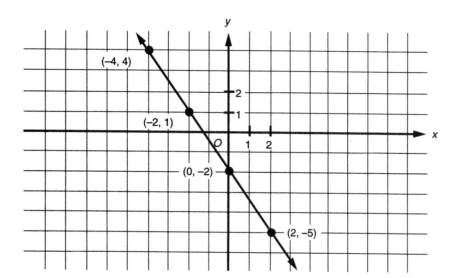

The $x$-intercept is $\dfrac{-4}{3}$.

The $y$-intercept is $-2$.

10.

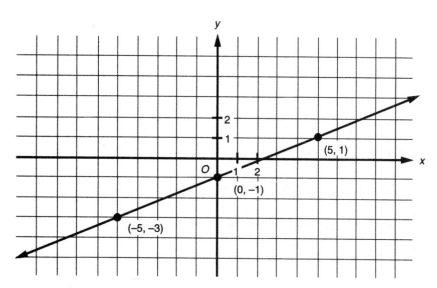

$$y = \frac{2}{5}x - 1$$

$$0 = \frac{2}{5}x - 1$$

$$0 = 2x - 5$$

$$-2x = -5$$

$$x = \frac{5}{2}$$

The $x$-intercept is $\dfrac{5}{2}$ or 2.5.

$$y = \frac{2}{5}(0) - 1$$

$$= -1$$

The $y$-intercept is $-1$.

C11.  $y = -3.14x + 7.92$

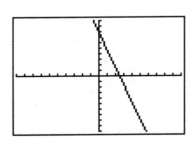

C12.  $y = 1.23x + 5.76$

a.  Press [Y=] [CLEAR] [1] [.] [2] [3] [X,T,θ,n] [+] [5] [.] [7] [6] [ENTER].

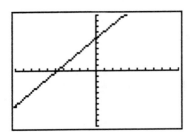

b.  To find the $y$-intercept, press [2nd] [CALC] [1] [0].
    The $y$-intercept is $y = 5.76$.

c.  To find the $x$-intercept, press [2nd] [CALC] [2].

    At the <u>Left Bound?</u> prompt, move the cursor to the left of the

    $x$-intercept and press [ENTER].

    At the <u>Right Bound?</u> prompt, move the cursor to the right of the

    $x$-intercept and press [ENTER].

    At the <u>Guess?</u> prompt, move the cursor between the two boundaries

    and press [ENTER].
    The $x$-intercept is −4.68.

d.  To find $y$, press [2nd] [CALC] [1] [(−)] [1] [.] [1] [2] [ENTER].
    The $y$-value is 4.3824.

# UNIT 24

1.

2.   $y = x^2 + 2x - 8$ with $a = 1$, $b = 2$, $c = -8$

Parabola opens up.

$y$-intercept is $-8$.

$x$-intercepts:   $0 = x^2 + 2x - 8$

$\qquad\qquad\qquad = (x + 4)(x - 2)$

$\qquad\qquad x = -4 \qquad$ or $\qquad x = 2$

$x$-intercepts are $-4$ and $2$.

Vertex:   $\dfrac{-b}{2a} = \dfrac{-2}{2} = -1$

If $x = -1$,

$\qquad y = (-1)^2 + 2(-1) - 8$

$\qquad\quad = 1 - 2 - 8$

$\qquad\quad = -9$.

Vertex at $(-1, -9)$.

3.

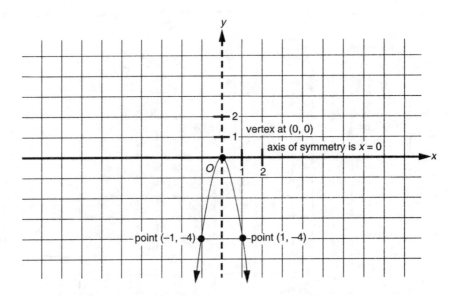

4.  $y = -x^2 - 4x - 3$ with $a = -1$, $b = -4$, $c = -3$

Parabola opens down.

$y$-intercept is $-3$.

$x$-intercepts:  $\begin{aligned} 0 &= -x^2 - 4x - 3 \\ &= x^2 + 4x + 3 \\ &= (x + 3)(x + 1) \end{aligned}$

$x = -3$    or    $x = -1$

$x$-intercepts are $-3$ and $-1$.

Vertex:    $x = -2$, midway between $x$-intercepts

$\begin{aligned} y &= -(-2)^2 - 4(-2) - 3 \\ &= -4 + 8 - 3 \\ &= 1 \end{aligned}$

Vertex at $(-2, 1)$.

No need to find point on either side of vertex because we have $x$-intercepts.

5.

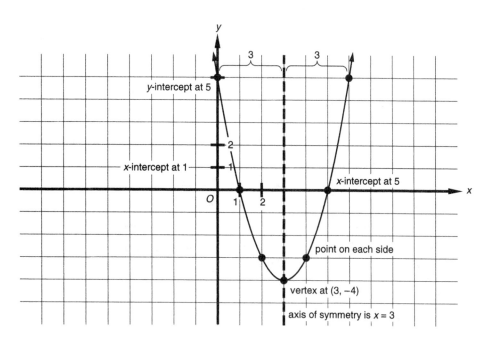

6.  $y = 4 + 2x^2$ with $a = 2$, $b = 0$, $c = 4$

Parabola opens up.

$y$-intercept is 4.

$x$-intercepts:  $0 = 4 + 2x^2$

$$-4 = 2x^2$$

No solution; there are no $x$-intercepts.

Vertex:  $x = \dfrac{-b}{2a} = \dfrac{-0}{2(2)} = 0$

$$y = 4 + 2(0)^2 = 4$$

Vertex at $(0, 4)$.

Find one point on either side:

let $x = 1$; they $y = 4 + 2(1)^2 = 4 + 2 = 6$  $(1, 6)$

let $x = -1$; then $y = 6$  $(-1, 6)$

7.

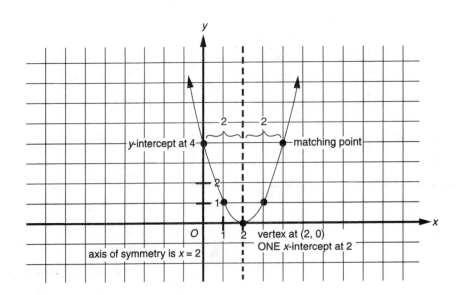

8. $y = 25 - x^2$ with $a = -1$, $b = 0$, $c = 25$

Parabola opens down.

$y$-intercept is 25.

$x$-intercepts:  $0 = 25 - x^2$

$\qquad\qquad\quad = (5 - x)(5 + x)$

$x$-intercepts are 5 and –5.

Vertex:  $\qquad x = 0$, midway between $x$-intercepts

$\qquad\qquad\quad y = 25 - 0^2 = 25$

$\qquad\qquad\quad$ Vertex at (0, 25).

Axis of symmetry is $x = 0$.

Find one point on either side:

let $x = 1$; then $y = 25 - 1^2 = 24$   (1, 24)

let $x = -1$; then $y = 24$ $\qquad$ (–1, 24)

9.

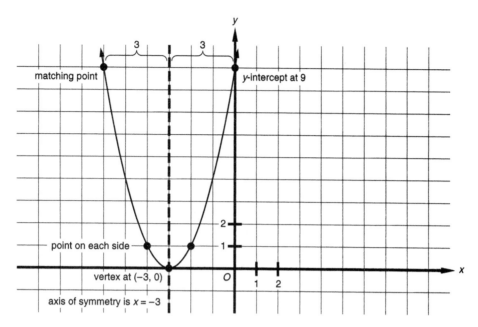

10. $y = 2x^2 + 4x - 1$ with $a = 2$, $b = 4$, $c = -1$

Parabola opens up.

$y$-intercept is $-1$.

$x$-intercepts: $0 = 2x^2 + 4x - 1$

Since the right-hand side is not easily factored, we will go to the next step.

Vertex:   $x = \dfrac{-b}{2a} = \dfrac{-4}{2(2)} = -1$

$y = 2(-1)^2 + 4(-1) - 1$

$= 2 - 4 - 1$

$= -3$

Vertex at $(-1, -3)$.

Find one point on either side:

let $x = 0$; then $y = 2(0)^2 + 4(0) - 1 = 0 + 0 - 1 = -1$     $(0, -1)$

let $x = -2$; then $y = 2(-2)^2 + 4(-2) - 1 = 8 - 8 - 1 = -1$     $(-2, -1)$

11.

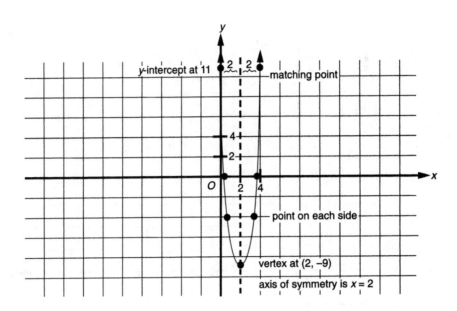

12. $y = -x^2 + 10x$ with $a = -1$, $b = 10$, $c = 0$

Parabola opens down.

$y$-intercept is 0.

$x$-intercepts: $0 = -x^2 + 10x$

$\phantom{x\text{-intercepts: }0} = -x(x - 10)$

$x$-intercepts are 0 and 10.

Vertex: $\quad x = 5$, midway between $x$-intercepts

$\phantom{Vertex:\quad} y = -25 + 10(5) = 25$

$\phantom{Vertex:\quad}$ Vertex at (5, 25)

13.

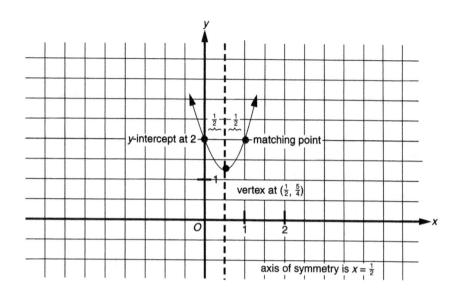

14. $y = 2x^2 - 12x + 3$ with $a = 2$, $b = -12$, $c = 3$

Parabola opens up.

$y$-intercept is 3.

Vertex: $\quad \dfrac{-b}{2a} = \dfrac{-(-12)}{2(2)} = \dfrac{12}{4} = 3$

$\qquad\qquad y = 2(9) - 12(3) + 3 = -15$

$\qquad\qquad$ Vertex at $(3, -15)$.

Find one point on either side:

let $x = 2$; then $y = 2(4) - 12(2) + 3 = -13$ $\quad$ $(2, -13)$

let $x = 4$; then $y = -13$ $\qquad\qquad\qquad$ $(4, -13)$

15.

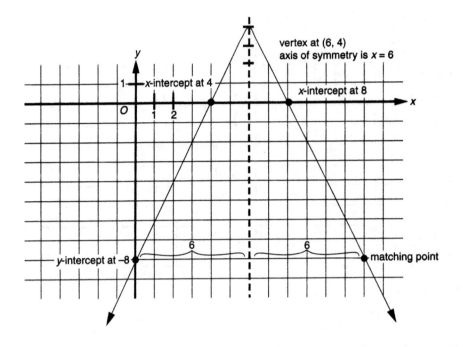

vertex at (6, 4)
axis of symmetry is $x = 6$

x-intercept at 4

x-intercept at 8

x

1

y-intercept at –8

6

6

matching point

## UNIT 25

Given:  $f(x) = x - 2$

1.   $f(7) = 5$

2.   $f(-3) = -3 - 2 = -5$

3.   $f(3.7) = 1.7$

4.   $f(0) = 0 - 2 = -2$

Given:  $g(x) = 3x$

5.   $g(2) = 6$

6.   $g(-5) = 3(-5) = -15$

7.   $g(0) = 0$

8.   $g(-4.1) = 3(-4.1) = -12.3$

Given:  $h(x) = 5x + 1$

9.   $h(10) = 51$

10.   $h(3) = 5(3) + 1 = 15 + 1 = 16$

11.   $h(0) = 1$

12.   $h(-4) = 5(-4) + 1 = -20 + 1 = -19$

Given:  $f(x) = 1 - x^2$

13.   $f(3) = -8$

14.   $f(-5) = 1 - (-5)^2 = 1 - 25 = -24$

15.   $f(-1) = 0$

Given:  $G = \{(5, 6), (7, 7), (8, 5), (6, 11)\}$

16. $D = \{5, 6, 7, 8\}$

17. $R = \{5, 6, 7, 11\}$

18. $G(8) = 5$

19. $G(5) = 6$

20. $G^{-1}(11) = 6$

21. $G^{-1}(5) = 8$

22. Yes

C23. $f(2.5) = 10.75$
$f(0.25) = 1.75$

C24. Given: $f(x) = 3x^4 + 2x^3 - x^2 + 1$

| Y= | CLEAR | 3 | X,T,θ,n | ∧ | 4 | + | 2 | X,T,θ,n | ∧ | 3 | − | X,T,θ,n | x² | + |

1 ENTER.

$f(2) = 61$
$f(1.5) = 20.688$

# UNIT 26

1. The solution to the system is $x = 4$ and $y = 1$.

2. $13x - 5y = -5$
$5(x + y = 1)$
$13x - 5y = -5$
$\underline{5x + 5y = 5}$
$18x \quad = 0$
$x = 0$

If $x = 0$ and $x + y = 1$,
$0 + y = 1$,
$y = 1$.

The solution to the system is $x = 0$ and $y = 1$.

3. The solution to the system is $x = -2$ and $y = -12$.

4. $+2(5x + 3y = 1$
$5(-2x + 5y = 12)$

$\phantom{+}10x + 6y = 2$
$\underline{-10x + 25y = 60}$
$\phantom{+1010x +}31y = 62$
$\phantom{+1010x +10}y = 2$

If $y = 2$ and $5x + 3y = 1$
$$5x + 3(2) = 1,$$
$$5x + 6 = 1,$$
$$5x = -5,$$
$$x = -1.$$

The solution to the system is $x = -1$ and $y = 2$.

5. The solution to the system is $x = 2$ and $y = 6$.

6. $2x + y = 0$
$\underline{x - y = 1}$
$3x \phantom{- y} = 1$

$$x = \frac{1}{3}$$

If $x = \dfrac{1}{3}$ and $x - y = 1$,

$$\frac{1}{3} - y = 1,$$

$$-y = \frac{2}{3},$$

$$y = -\frac{2}{3}.$$

The solution to the system is $x = \dfrac{1}{3}$ and $y = -\dfrac{2}{3}$.

7. The solution to the system is $x = 0$ and $y = 2$.

8. $11(-2x + 17y = 6)$
$2(11x - 5y = -33)$

$-22x + 187y = 66$
$\underline{22x - 10y = -66}$
$\phantom{-22x +1}177y = 0$
$\phantom{-22x +10}y = 0$

If $y = 0$ and $-2x + 17y = 6$,
$$-2x + 0 = 6,$$
$$x = -3.$$

The solution to the system is $x = -3$ and $y = 0$.

9.  The solution to the system is $x = 2$ and $y = 20$.

10.  $$\begin{aligned} 4x + \quad y &= 13 \\ -4(x - \quad 3y &= \quad 0) \end{aligned}$$

$$\begin{aligned} 4x + \quad y &= 13 \\ -4x + 12y &= \quad 0 \\ \hline 13y &= 13 \\ y &= \quad 1 \end{aligned}$$

If $y = 1$ and $x - 3y = 0$,

$$x - 3(1) = 0,$$

$$x = 3.$$

The solution to the system is $x = 3$ and $y = 1$.

11.  The solution to the system is $x = 2$ and $y = 7$.

12.  $$x + 4y \qquad = 8$$

$$y = -\frac{1}{4}x - 7$$

$$x + 4y \qquad = 8$$

$$x + 4\left(-\frac{1}{4}x - 7\right) = 8 \quad \text{by substitution}$$

$$x - x \qquad -28 = 8$$

$$-28 = 8$$

There is no solution to the system.

13.  The solution to the system is $x = 5$ and $y = 1$.

14.  $$2\left(2x + \frac{1}{2}y = 2\right)$$

$$6x - y = 1$$

$$\begin{aligned} 4x + y &= 4 \\ 6x - y &= 1 \\ \hline 10x \quad &= 5 \end{aligned}$$

$$x = \frac{1}{2}$$

If $x = \frac{1}{2}$ and $6x - y = 1$,

$$6\left(\frac{1}{2}\right) - y = 1,$$

$$3 - y = 1,$$

$$2 = y.$$

The solution to the system is $x = \frac{1}{2}$ and $y = 2$.

15. The solution to the system is $x = 1$ and $y = 12$.

16. $5\left(y = -\dfrac{2}{5}x + 4\right)$

$5y = -2x + 20$

$2x + 5y = 20$

$3\left(x = \dfrac{1}{3}y - 7\right)$

$3x = y - 21$

$3x - y = -21$

$2x + 5y = 20$

$5(3x - y = -21)$

$2x + 5y = 20$

$\underline{15x - 5y = -105}$

$17x = -85$

$x = -5$

If $x = -5$ and $y = -\dfrac{2}{5}x + 4$,

$$y = -\dfrac{2}{5}(-5) + 4,$$

$$= 2 + 4,$$

$$= 6.$$

The solution to the system is $x = -5$ and $y = 6$.

17. There is no solution to the system.

18. $11x - y = 10$

$\underline{2x + y = 7}$

$13x = 17$

$$x = \dfrac{17}{13}$$

If $x = \dfrac{17}{13}$ and $2x + y = 7$,

$$2\left(\dfrac{17}{13}\right) + y = 7,$$

$$\dfrac{34}{13} + y = 7.$$

Clear of fractions: $34 + 13y = 91$,

$$13y = 57$$

$$y = \dfrac{57}{13}$$

The solution to the system is $x = \dfrac{17}{13}$ and $y = \dfrac{57}{13}$.

19. The solution to the system is $a = 2$ and $b = 2$.

20. $2x = -7(y + 1)$
    $2x = -7y - 7$
    $2x + 7y = -7$

    Use substitution, since $y = -\frac{2}{7}x - 1$:

    $2x + 7y \qquad\qquad = -7$

    $2x + 7\left(-\frac{2}{7}x - 1\right) = -7$

    $2x - 2x \qquad -7 = -7$

    $\qquad\qquad\qquad -7 = -7$

    There are infinitely many solutions to the system.

## UNIT 27

1. Use $x = 2$ to substitute:

    $y = 3x^2 - 7x + 11$
    $\phantom{y} = 3(4) - 7(2) + 11$
    $\phantom{y} = 12 - 14 + 11$
    $\phantom{y} = 9$

    The solution to the system is $x = 2$ and $y = 9$ or $(2, 9)$.

2. Use $y = 4x^2$ to substitute:

    $-4x + y = 0$

    $-4x + 4x^2 = 0$
    $4x^2 - 4x = 0$
    $-4x(x - 1) = 0$

    $\qquad 4x = 0 \qquad\qquad x - 1 = 0$
    $\qquad\phantom{4} x = 0 \qquad\qquad\phantom{x - 1} x = 1$

    If $\qquad x = 0 \qquad$ If $\qquad x = 1$
    and $\qquad y = 4x^2, \qquad$ and $\quad y = 4x^2,$
    $\qquad\qquad y = 4(0) \qquad\qquad\quad y = 4(1)$
    $\qquad\qquad\phantom{y} = 0. \qquad\qquad\qquad\phantom{y} = 4.$

    There are two solutions to the system.
    $x = 0$ and $y = 0 \qquad\qquad (0, 0)$
    and
    $x = 1$ and $y = 4 \qquad\qquad (1, 4)$

3. There are two solutions to the system:

   $x = 4$ and $y = 36$      (4, 36)

   and

   $x = -11$ and $y = 66$      (−11, 66)

4. Use $y = -x^2 - 1$ to substitute:

   $$-x + 5y = -5$$
   $$-x + 5(-x^2 - 1) = -5$$
   $$-x - 5x^2 - 5 = -5$$
   $$-5x^2 - x = 0$$
   $$5x^2 + x = 0$$
   $$x(5x + 1) = 0$$

   $x = 0$      $5x + 1 = 0$

   $$5x = -1$$
   $$x = -\frac{1}{5}$$

   If $x = 0$        If $x = -\frac{1}{5}$,

   $y = -x^2 - 1$      $y = -x^2 - 1$

   $= 0 - 1$         $= -\left(-\dfrac{1}{5}\right)^2 - 1$

   $= -1.$          $= -\dfrac{1}{25} - 1$

                   $= -\dfrac{26}{25}.$

   There are two solutions to the system:

   $x = 0$ and $y = -1$      (0, −1)

   and

   $x = -\dfrac{1}{5}$ and $y = -\dfrac{26}{25}$      $\left(-\dfrac{1}{5}, -\dfrac{26}{25}\right)$

5. There are two solutions to the system:

   $x = 3$ and $y = 3$      (3, 3)

   and

   $x = -7$ and $y = -7$      (−7, −7)

6. Use $y = x^2 + 2x + 5$ to substitute:

   $$-4x + 2y = -7$$

   $$-4x + 2(x^2 + 2x + 5) = -7$$
   $$-4x + 2x^2 + 4x + 10 = -7$$
   $$2x^2 = -17$$

   There is no solution to the system.

7. The solution to the system is $x = 3$ and $y = 6$ or (3, 6).

8. Use $y = 1$ to substitute:

$$y = 3x^2 + 8x - 2$$

$$\downarrow$$

$$1 = 3x^2 + 8x - 2$$
$$0 = 3x^2 + 8x - 3$$
$$0 = (3x - 1)(x + 3)$$

| | |
|---|---|
| $3x - 1 = 0$ | $x + 3 = 0$ |
| $3x = 1$ | $x = -3$ |
| $x = \dfrac{1}{3}$ | |

There are two solutions to the system.

$x = \dfrac{1}{3}$ and $y = 1$  $\left(\dfrac{1}{3}, 1\right)$

and

$x = -3$ and $y = 1$  $(-3, 1)$

9. The solution to the system is $x = -5$ and $y = 38$ or (-5, 38).

10. Use $y = x^2 + 5x + 1$ to substitute:

$$y = x^2 + 3x + 9$$

$$x^2 + 5x + 1 = x^2 + 3x + 9$$
$$2x = 8$$
$$x = 4$$

If $x = 4$ and $y = x^2 + 5x + 1$,

$$y = (4)^2 + 5(4) + 1,$$
$$= 16 + 20 + 1,$$
$$= 37.$$

The solution to the system is $x = 4$ and $y = 37$ or (4, 37).

11. The solution to the system is $x \approx 1.127$ and $y \approx -2.365$ or (1.127, -2.365).

## UNIT 28

1. $x \leq 1$

2. $4 - (12 - 3x) \le -5$

$\qquad 4 - 12 - 3x \le -5$

$\qquad -8 + 3x \le -5$

$\qquad 3x \le -5 + 8$

$\qquad 3x \le 3$

$\qquad x \le 1$

3. $x < 3$

4. $4x + (3x - 7) > 2x - (28 - 2x)$

$\qquad 4x + 3x - 7 > 2x - 28 + 2x$

$\qquad 7x - 7 > 4x - 28$

$\qquad 3x > -21$

$\qquad x > -7$

$\qquad -7 < x$

5. $x \ge -6$

6. $3x + 4(x - 2) \ge x - 5 + 3(2x - 1)$

$\qquad 3x + 4x - 8 \ge x - 5 + 6x - 3$

$\qquad 7x - 8 \ge 7x - 8$

$\qquad 0 \ge 0$

The solution is the entire set of real numbers.

7. $x < \dfrac{-10}{3}$.

8. $3x - 2(x - 5) < 3(x - 1) - 2x - 11$

$3x - 2x + 10 < 3x - 3 - 2x - 11$

$x + 10 < x - 14$

$x - x < -14 - 10$

$0 < -24$

There is no solution.

9. The solution is the entire set of real numbers.

10. $5x - 2(3x - 4) > 4[2x - 3(1 - 3x)]$

$5x - 6x + 8 > 4[2x - 3 + 9x]$

$-x + 8 > 8x - 12 + 36x$

$-x + 8 > 44x - 12$

$-x - 44x > -12 - 8$

$-45x > -20$

$x < \dfrac{20}{45}$ or $\dfrac{4}{9}$

# UNIT 29

The details for graphing many of the following parabolas can be found in the answers to the exercises for Unit 24.

1. $-7 < x < 3$

2. A rough sketch of $y = x^2 + 2x - 8$ is shown at the right. For what values of $x$ is the curve **below** the $x$-axis?

   $-4 < x < 2$

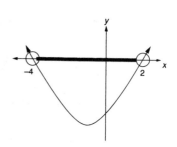

3. $x < -3$ or $4 < x$

4. A rough sketch of $y = -x^2 - 4x - 3$ is shown at the right. For what values of $x$ is the curve **below** the $x$-axis?

   $x < -3$ or $-1 < x$

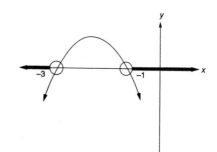

5. $-2 < x < 1$

6. A rough sketch of $y = -4x^2$ is shown at the right. For what values of $x$ is the curve **above** or **on** the $x$-axis?

   $x = 0$

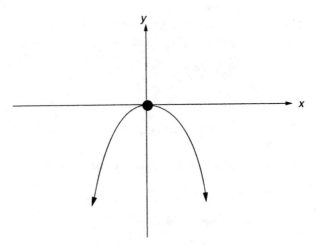

7. $x = 2$

8. A rough sketch of $y = 4 + 2x^2$ is shown at the right. For what values of $x$ is the curve **below** or **on** the $x$-axis?
   There is no solution.

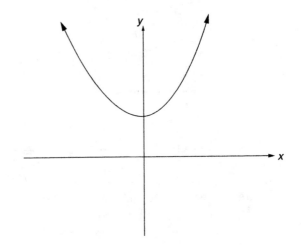

9. $0 \leq x \leq 1$

10. A sketch of $y = 25 - x^2$ is shown at the right. For what values of $x$ is the curve **above** or **on** the $x$-axis?

    $-5 \leq x \leq 5$

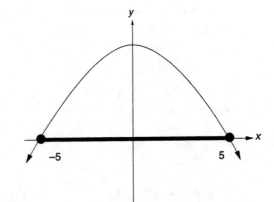

11. The solution is the set of all real numbers.

12. A sketch of $y = x^2 + 6x + 9$ is shown at the right. For what values of $x$ is the curve **above** the x-axis?

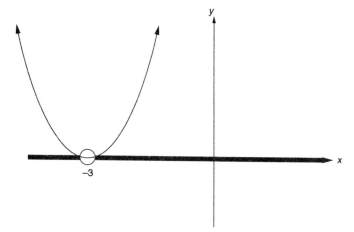

$x < -3$ or $-3 < x$ or, in other words, all reals except $-3$

13. $x \le -3$ or $5 \le x$

14. A sketch of $y = 2x^2 + 4x - 1$ is shown at the right. For what values of $x$ is the curve **below** the x-axis?

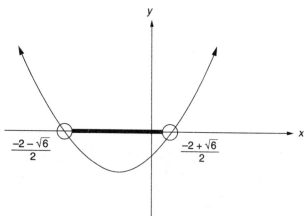

$$\frac{-2 - \sqrt{6}}{2} < x < \frac{-2 + \sqrt{6}}{2}$$

15. There is no solution.

16. A sketch of $y = -x^2 + 10x$ is shown at the right. For what values of $x$ is the curve **below** or **on** the $x$-axis?

$x \le 0$ or $10 \le x$

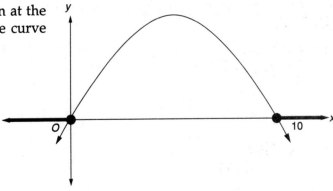

17. $-\dfrac{5}{2} \le x \le 1$

18. A sketch of $y = 3x^2 - 3x + 2$ is shown at the right. For what values of $x$ is the curve **above** the $x$-axis?

The solution is the set of all real numbers.

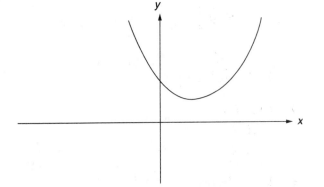

19. $-\dfrac{2}{3} \le x \le \dfrac{1}{2}$

20. A sketch of $y = 2x^2 - 12x + 3$ is shown at the right. For what values of $x$ is the curve **below** or **on** the $x$-axis?

$\dfrac{12 - \sqrt{120}}{4} \le x \le \dfrac{12 + \sqrt{120}}{4}$

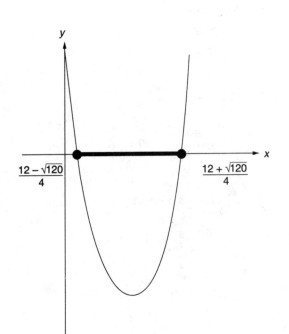

## UNIT 30

1. $\log_3 81 = x$
   $3^x = 81$
   $\quad = 3^4$
   $x = 4$

2. $\log_5 125 = x$
   $5^x = 125$
   $\quad = 5^3$
   $x = 3$

3. $\log_7\left(\dfrac{1}{7}\right) = x$
   $7^x = \dfrac{1}{7}$
   $\quad = 7^{-1}$
   $x = -1$

4. $\log 1 = x$
   $10^x = 1$
   $x = 0$

5. $\log_3 x = 2$
   $3^2 = x$
   $9 = x$

6. $\log_7 x = 0$
   $7^0 = x$
   $1 = x$

7. $\log_9 x = \dfrac{1}{2}$
   $9^{1/2} = x$
   $\sqrt{9} = x$
   $3 = x$

8. $\log_x 27 = 3$
   $x^3 = 27$
   $\quad = 3^3$
   $x = 3$

9. $\log_x 49 = 2$
   $x^2 = 49$
   $\quad = 7^2$
   $x = 7$

10. $\log_x 121 = 2$
    $x^2 = 121$
    $x = 11$

11. $\log x^5 = 5\log x$

12. $\log 2xy^3$
    $\log 2 + \log x + \log y^3$
    $\log 2 + \log x + 3\log y$

13. $\log \dfrac{x^2}{y}$
    $\log x^2 - \log y$
    $2\log x - \log y$

14. $\log \dfrac{x}{yz}$
    $\log x - \log yz$
    $\log x - (\log y + \log z)$
    $\log x - \log y - \log z$

15. $\log \sqrt{x^3 y}$
    $\log(x^3 y)^{1/2}$
    $\dfrac{1}{2}\log x^3 y$
    $\dfrac{1}{2}(\log x^3 + \log y)$
    $\dfrac{1}{2}(3\log x + \log y)$
    $\dfrac{3}{2}\log x + \dfrac{1}{2}\log y$

## UNIT 31

1.

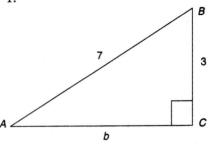

$$c^2 = a^2 + b^2$$
$$7^2 = 3^2 + b^2$$
$$49 = 9 + b^2$$
$$40 = b^2$$
$$b^2 = 40$$
$$b = \sqrt{40}$$
$$\phantom{b} = 2\sqrt{10}$$

2.

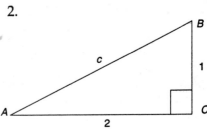

$$c^2 = a^2 + b^2$$
$$\phantom{c^2} = 1^2 + 2^2$$
$$\phantom{c^2} = 1 + 4$$
$$\phantom{c^2} = 5$$
$$c = \sqrt{5}$$

3.

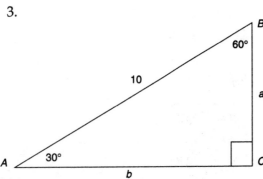

$$a = \frac{c}{2}$$
$$\phantom{a} = \frac{10}{2}$$
$$\phantom{a} = 5$$
$$b = \frac{c}{2}\sqrt{3}$$
$$\phantom{b} = \frac{10}{2}\sqrt{3}$$
$$\phantom{b} = 5\sqrt{3}$$

4.

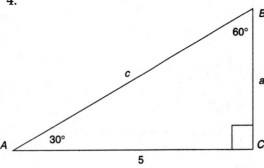

$$b = \frac{c}{2}\sqrt{3}$$
$$5 = \frac{c}{2}\sqrt{3}$$
$$10 = c\sqrt{3}$$
$$\frac{10}{\sqrt{3}} = c$$
$$a = \frac{c}{2}$$
$$\phantom{a} = \frac{10}{\sqrt{3}} \cdot \frac{1}{2}$$
$$\phantom{a} = \frac{5}{\sqrt{3}}$$

5.

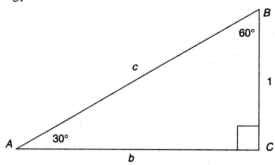

$$a = \frac{c}{2}$$

$$1 = \frac{c}{2}$$

$$2 = c$$

$$b = \frac{c}{2}\sqrt{3}$$

$$= \frac{2}{2}\sqrt{3}$$

$$= \sqrt{3}$$

6.

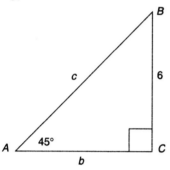

$$a = b$$

$$6 = b$$

$$a = \frac{c}{2}\sqrt{2}$$

$$6 = \frac{c}{2}\sqrt{2}$$

$$12 = c\sqrt{2}$$

$$\frac{12}{\sqrt{2}} = c$$

7.

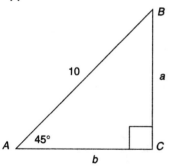

$$a = \frac{c}{2}\sqrt{2}$$

$$= \frac{10}{2}\sqrt{2}$$

$$= 5\sqrt{2}$$

$$b = 5\sqrt{2}$$

8.

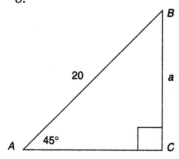

$$a = \frac{c}{2}\sqrt{2}$$

$$= \frac{20}{2}\sqrt{2}$$

$$= 10\sqrt{2}$$

The ladder reaches up $10\sqrt{2}$ ft.

9.

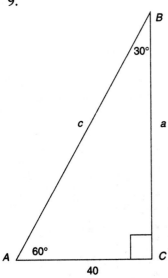

$c = 80$

$a = \dfrac{c}{2}\sqrt{3}$

$= \dfrac{80}{2}\sqrt{3}$

$= 40\sqrt{3}$

The building is $40\sqrt{3}$ ft high.

10.

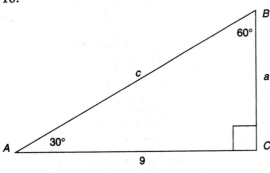

$b = \dfrac{c}{2}\sqrt{3}$

$9 = \dfrac{c}{2}\sqrt{3}$

$18 = c\sqrt{3}$

$\dfrac{18}{\sqrt{3}} = c$

$a = \dfrac{c}{2}$

$= \dfrac{18}{\sqrt{3}} \cdot \dfrac{1}{2}$

$= \dfrac{9}{\sqrt{3}}$

The observer is $\dfrac{9}{\sqrt{3}}$ ft tall.

# Index